ECONOMIC ADAPTATION

About the Authors

Prof. (Dr.) Udai Prakash Sinha has devoted more than 45 years of his career serving prominent organizations and universities of India and delivered the best of his capabilities fulfiling the need of the organizations and meeting their standard to take its pride and honor to the next level. In the span of 45 years, Prof. Sinha authored 24 books and more than 125 research papers from various International and National publications and journals. As a belonging of Economics, he has served the Economics department of several Institutions and successfully delivered more than 55 PhDs under his supervision.

Professor Sinha also served as a Vice Chancellor of Central University of Haryana and additionally as Executive Council too. Expert members of different Committee of UGC, Chairman and members of NAAC Committee of various colleges/Universities come under Indian Government, Expert of Economics at CSTT (Commission for Scientific and Technical Terminology, Government of India.

Dr. Kavitha S. Sharma serves as an Assistant Professor at the Faculty of Management Studies (FMS), University of Delhi. With a strong academic and professional background in accounting and finance. Before transitioning into academia, Dr. Sharma worked as a financial and commodities analyst, which enriched her insights into diverse market dynamics and financial instruments.

She holds an MBA from Anna University and an M.Com from Guru Jambheshwar University, Hisar. She was awarded her Ph.D. in Management by T.M. Bhagalpur University. Dr. Sharma has published several research papers in reputed national and international journals and has undertaken various research projects in her domain.

An active contributor to academic discourse, Dr. Sharma has presented her work at numerous national and international conferences. She has also been invited as a keynote speaker at several academic seminars. She is a Life Member of the Indian Accounting Association, South Haryana Chapter, reflecting her ongoing engagement with the broader accounting and finance community.

ECONOMIC ADAPTATION

Responding to Climate Change Threats

Udai Prakash Sinha
Kavitha S. Sharma

CONCEPT PUBLISHING COMPANY PVT. LTD.
NEW DELHI - 110 059

ISBN: 978-93-6344-479-9

First Published 2025

Published and Printed by

Concept Since 1974

Concept Publishing Company Pvt. Ltd.
A/15-16, Commercial Block, Mohan Garden
New Delhi-110059 (India)
T: +91-11-41101460
E: publishing@conceptpub.com, *W*: www.conceptpub.com

Cataloging in Publication--Courtesy: D.K. Agencies (P) Ltd. <docinfo@dkagencies.com>

Sinha, Udai Prakash, 1956- **author.**
Economic adaptation : responding to climate change threats / Udai Prakash Sinha, Kavitha S. Sharma.
pages cm
Includes bibliographical references and index.
ISBN 9789363444799

1. Climatic changes--Economic aspects. 2. Climate change adaptation--Economic aspects. 3. Climatic changes--International cooperation. 4. Sustainable development--International cooperation. 5. Climatic changes--Government policy. I. Sharma, Kavitha S., author. II. Title.

LCC QC903.S56 2025 | DDC 363.73874 23

Preface

In recent decades, the world has witnessed an undeniable shift in climate patterns, with profound implications for economies, ecosystems, and societies worldwide. As temperatures rise, weather patterns hecome increasingly unpredictable, and the frequency and intensity of extreme weather events escalate, the need for proactive economic adaptation to climate change has never been more urgent. "Economic Adaptation: Responding to Climate Change Threats" delves into the complex intersection of economics and climate change, exploring the multifaceted challenges posed by a warming world and offering insights into innovative strategies for building resilience and fostering sustainable economic growth.

This book represents a collaborative effort by leading experts and scholars in the fields of economics, environmental science, policy, and business. Through a comprehensive examination of key concepts, case studies, and policy frameworks, it aims to provide readers with a holistic understanding of the economic dimensions of climate change adaptation. This book delves into the role of policy frameworks, investment strategies, and technological innovations in enhancing economic resilience and fostering sustainable development pathways. It examines the economic implications of climate-related risks on key sectors such as agriculture, energy, infrastructure, and finance, offering practical insights into how businesses and policymakers can navigate the challenges and seize the opportunities presented by a changing climate.

At its core, "Economic Adaptation" is a call to action to transcend traditional boundaries and embrace bold, interdisciplinary approaches to addressing one of the defining challenges of our time. It is a testament to the power of human ingenuity and collaboration in overcoming adversity and building a more resilient, prosperous future for generations to come. Together, let us embark on a journey of economic adaptation, guided by the principles of stewardship, innovation, and solidarity, as we strive to safeguard our planet and ensure prosperity for all.

Udai Prakash Sinha
Kavitha S. Sharma

Contents

Preface v

1. Understanding Climate Change Economics: Challenges and Opportunities 1
2. Historical Perspectives on Economic Adaptation to Climate Change 46
3. The Economics of Climate Risk Assessment and Management 86
4. Policy Frameworks for Economic Resilience in a Changing Climate 126
5. Investing in Climate-Resilient Infrastructure and Innovation 159
6. Agriculture and Food Security in a Warming World: Economic Implications 187
7. Energy Transition Economics: Shifting Towards Renewable Resources 213
8. Coastal Cities and Economic Adaptation: Mitigating Risks from Rising Seas 256
9. Economic Impacts of Extreme Weather Events: Lessons Learned and Future Preparedness 288
10. Sustainable Business Practices: Aligning Profit with Planet 323

11. Social Justice and Economic Adaptation: Addressing Inequities 353

12. Future Directions for Economic Adaptation to Climate Change 367

Bibliography 377

Index 383

1

Understanding Climate Change Economics

Challenges and Opportunities

Introduction to Climate Change Economics

Climate change economics is an interdisciplinary field that examines the economic impacts of climate change and the cost-effectiveness of various adaptation and mitigation strategies. It involves understanding how climate change affects economic systems, the distribution of these impacts across different sectors and regions, and the policy measures necessary to address these challenges. This chapter provides a foundational understanding of climate change economics, highlighting the critical importance of economic adaptation in the face of ongoing environmental changes. One of the most significant challenges in economic adaptation to climate change is the pervasive uncertainty and risk associated with climate variability. The complexity of climate systems makes it difficult to predict the exact impacts of climate change, leading to substantial uncertainties that affect both short-term and long-term economic planning. This uncertainty stems from various factors, including the inherent variability of the climate, the unpredictability of future greenhouse gas emissions, and the limitations of current climate models.

Climate Variability

Climate variability refers to the natural fluctuations in climate patterns over different time scales, from years to decades. These fluctuations can significantly influence weather events, such as the frequency and intensity of storms, droughts, and heat waves. The challenge lies in differentiating between natural climate variability and changes induced by human activities. Predicting the exact impacts of climate change on economic systems is further complicated by this natural variability. For instance, while climate models can provide projections of future climate conditions, these models are based on various assumptions and scenarios regarding future emissions, technological advancements, and policy decisions. The Intergovernmental Panel on Climate Change (IPCC) uses multiple scenarios to account for different potential futures, but even the most sophisticated models cannot capture all the nuances of the climate system. This results in a range of possible outcomes, each with different implications for economic adaptation.

Economic Uncertainty

Climate change can affect economic systems in myriad ways, including altering agricultural productivity, disrupting supply chains, and increasing the frequency and severity of extreme weather events. Each of these impacts carries its own set of uncertainties. Agriculture, for example, is highly sensitive to changes in temperature, precipitation patterns, and the frequency of extreme weather events. While some regions might experience increased crop yields due to longer growing seasons, others might suffer from reduced yields due to droughts or heat waves. This variability creates economic uncertainty for farmers, agribusinesses, and policymakers.

Similarly, supply chains can be disrupted by climate-related events such as floods, hurricanes, and wildfires, leading to economic losses and increased costs for businesses.

Risk Management and Adaptation Strategies

Effective risk management is crucial for economic adaptation to climate change. This involves identifying, assessing, and prioritizing risks, as well as implementing strategies to mitigate these risks. However, the uncertainty associated with climate change makes risk management particularly challenging. One approach to managing climate risks is through the development of robust and flexible adaptation strategies. These strategies are designed to be effective under a wide range of future climate conditions, rather than relying on precise predictions. For example, infrastructure projects can be designed to withstand a range of potential climate impacts, such as higher temperatures and increased precipitation. By building flexibility into adaptation measures, governments and businesses can reduce the risk of maladaptation—investing in measures that prove to be inadequate or inappropriate under future conditions.

The Role of Insurance and Financial Markets

Insurance and financial markets play a critical role in managing the economic risks associated with climate change. Insurance products can help businesses and individuals recover from climate-related losses, while financial markets can provide the necessary capital for investment in adaptation measures. However, the insurance industry faces its own set of challenges in the context of climate change.

As the frequency and severity of climate-related disasters increase, insurers may experience higher claims, leading to increased premiums and potential financial instability. To address these challenges, the insurance industry is developing new products and pricing models that better account for climate risks. For example, parametric insurance, which pays out based on the occurrence of predefined events (such as a certain level of rainfall), is being increasingly used to provide quicker and more predictable payouts. Financial markets also have a role to play in supporting economic adaptation. Green bonds and other climate-related financial instruments can raise funds for projects that enhance climate resilience, such as renewable energy installations and climate-resilient infrastructure. By channelling investment into these areas, financial markets can help reduce the economic risks associated with climate change.

Government and Policy Responses

Governments play a crucial role in managing the economic risks of climate change through policy and regulation. Effective policies can incentivize businesses and individuals to adopt adaptive measures, reduce greenhouse gas emissions, and build resilience to climate impacts. However, the uncertainty associated with climate change complicates policy-making. Policies must be designed to be flexible and adaptive, allowing for adjustments as new information becomes available. For example, governments can implement adaptive management frameworks that allow for periodic review and revision of policies based on the latest scientific and economic data. Additionally, international cooperation and coordination are essential, as climate change is a global challenge that requires collective action.

Conclusion

The uncertainty and risk associated with climate variability present significant challenges for economic adaptation to climate change. Predicting the exact impacts of climate change remains difficult due to the complex nature of climate systems and the limitations of current models. This uncertainty affects economic planning and decision-making, making it essential to develop robust and flexible adaptation strategies. Effective risk management, supported by insurance and financial markets and adaptive policy frameworks are crucial for building economic resilience to climate change. By addressing these challenges, societies can better prepare for the uncertain future posed by climate change.

Economic Uncertainty

The economic impacts of climate change are inherently uncertain, presenting significant challenges for businesses, governments, and policymakers in their efforts to plan and allocate resources effectively. This uncertainty arises from several interrelated factors, including the unpredictable nature of climate change, the complex interactions between climate and economic systems, and the varying responses of different economic sectors and regions to climate-related disruptions. Understanding and managing this economic uncertainty is crucial for developing resilient and adaptive strategies that can mitigate the adverse effects of climate change.

Unpredictable Nature of Climate Change

Climate change is characterized by a high degree of unpredictability, both in terms of the specific environmental changes that will occur and their timing. While scientific models provide a range of potential future scenarios, these projections come with significant uncertainties. For instance, predicting the precise changes in temperature, precipitation patterns, and the frequency of extreme weather events at regional and local levels remains challenging. This unpredictability complicates economic planning as businesses and governments struggle to prepare for a wide array of possible future conditions. One prominent example of this unpredictability is the impact of climate change on agriculture. Changes in temperature and precipitation can alter growing seasons, affect crop yields, and increase the prevalence of pests and diseases. However, these impacts can vary widely across different regions and crops. Farmers and agribusinesses must make decisions about planting, harvesting, and investing in infrastructure without knowing exactly how climate change will affect their specific operations. This uncertainty can lead to either underpreparation, resulting in significant losses, or overpreparation, which can be costly and inefficient.

Complex Interactions between Climate and Economic Systems

The interactions between climate change and economic systems are complex and multifaceted, further contributing to economic uncertainty. Climate change can affect various sectors of the economy in different ways, creating a web of interdependencies and feedback loops. For example, extreme weather events can disrupt supply chains, damage infrastructure, and reduce labour productivity, leading to cascading effects across multiple industries. Take the energy sector, for instance. Climate change can influence energy demand and supply in various ways.

Higher temperatures may increase the demand for cooling in the summer, while milder winters might reduce the need for heating. At the same time, extreme weather events such as hurricanes and floods can damage energy infrastructure, leading to supply disruptions and higher costs. Energy companies must navigate these uncertainties while making long-term investment decisions about infrastructure, technology, and resource allocation.

Sectoral and Regional Variability

The economic impacts of climate change are not uniform; they vary significantly across different sectors and regions. This variability adds another layer of uncertainty to economic planning and resource allocation. Some sectors, like agriculture and tourism, are highly sensitive to climate change, while others like information technology might be less directly affected but still face indirect consequences through supply chain disruptions or changes in consumer behaviour. Similarly, the regional impacts of climate change can be highly uneven. Coastal areas might face heightened risks from sea-level rise and increased storm surges, while inland regions could experience more severe droughts or heat waves. These regional differences necessitate tailored adaptation strategies that address specific local conditions, but they also complicate national and global economic planning.

Difficulty in Long-term Planning

Long-term economic planning is particularly challenging in the face of climate change due to the inherent uncertainty in predicting future conditions and the long time horizons involved. Infrastructure investments, for example, often have lifespan of several decades. Deciding where and how to build infrastructure that will remain resilient to climate impacts over its entire lifespan requires careful consideration of uncertain future climate conditions. Governments and businesses must weigh the costs and benefits of different adaptation strategies under various scenarios. For instance, investing in climate-resilient infrastructure may involve higher upfront costs but can provide significant long-term benefits by reducing the risks of climate-related damage and disruptions. However, determining the appropriate level of investment is difficult given the uncertainties surrounding future climate impacts and the potential for unexpected changes in technology, policy, and market conditions.

The Role of Policy and Regulation

Effective policy and regulation are essential for managing the economic uncertainties associated with climate change.

Governments can play a crucial role by providing clear and consistent policy frameworks that guide businesses and investors in their decision-making processes. Policies that promote research and development incentivize the adoption of climate-resilient technologies, and support vulnerable sectors and communities can help mitigate some of the economic uncertainties. However, policymaking itself is subject to uncertainties and political dynamics. Changes in government priorities, political will, and international agreements can all influence the direction and effectiveness of climate-related policies. Policymakers must, therefore, design flexible and adaptive policies that can be adjusted as new information becomes available and as circumstances change.

Conclusion

The economic impacts of climate change are fraught with uncertainty, posing significant challenges for businesses, governments, and policymakers in their efforts to plan and allocate resources effectively. The unpredictable nature of climate change, the complex interactions between climate and economic systems, and the sectoral and regional variability of impacts, all contribute to this uncertainty. Long-term planning is particularly difficult, requiring careful consideration of a wide range of possible future scenarios. Effective policy and regulation, along with robust risk management and adaptation strategies, are essential for navigating these uncertainties and building resilient economic systems capable of withstanding the impacts of climate change. By addressing these challenges, societies can better prepare for the uncertain economic landscape shaped by a changing climate.

Resource Allocation

Resource allocation is a critical component of economic adaptation to climate change, as it involves directing financial, human, and technological resources towards initiatives that enhance resilience and mitigate adverse impacts. However, several significant challenges complicate effective resource allocation, particularly funding constraints and competing priorities.

Funding Constraints

Securing sufficient funding for adaptation projects is one of the most pressing challenges, especially for developing countries. These nations often lack the financial resources required to implement large-scale adaptation measures, such as building climate-resilient infrastructure, developing early warning systems, and supporting vulnerable communities.

Even in wealthier nations, competing demands for public funds can limit the availability of resources for climate adaptation. One of the primary reasons for funding constraints is the high upfront costs associated with many adaptation projects. For example, constructing flood defences, upgrading infrastructure to withstand extreme weather events, or transitioning to climate-smart agricultural practices requires significant initial investment. While these projects can yield substantial long-term benefits by reducing the risks and costs of climate impacts, the initial financial outlay can be prohibitive, particularly for countries with limited budgets.

Moreover, developing countries often face additional financial challenges, such as high levels of debt, limited access to international finance, and economic instability. These factors can make it difficult to prioritize and secure funding for climate adaptation, even when the need is urgent. International financial mechanisms, such as the Green Climate Fund, aim to address these gaps by providing financial support to developing countries for adaptation and mitigation efforts. However, the scale of funding available through these mechanisms is often insufficient to meet the growing demands.

Competing Priorities

Balancing immediate economic needs with long-term adaptation goals is another significant challenge in resource allocation. Governments and businesses must make difficult decisions about where to allocate limited resources, often weighing the urgent demands of current economic and social issues against the imperative to invest in long-term climate resilience. Immediate economic needs, such as job creation, poverty alleviation, and infrastructure development, typically dominate policy agendas, especially in developing countries. These priorities can overshadow the need for adaptation investments, which may not yield immediate visible benefits.

For instance, a government may prioritize building new schools and hospitals over investing in climate-resilient infrastructure, even though the latter is crucial for safeguarding the former against future climate impacts. Additionally, political cycles and the short-term focus of electoral politics can exacerbate this issue. Elected officials may prioritize projects that deliver quick, tangible results to secure political support, rather than committing to long-term adaptation strategies that extend beyond their terms in office. This short-termism can hinder the implementation of comprehensive and effective adaptation plans.

Innovative Financing Mechanisms

To address funding constraints and competing priorities, innovative financing mechanisms are being explored and implemented. These mechanisms aim to leverage public and private sector resources, create new financial instruments, and foster partnerships to support climate adaptation. One such mechanism is the development of green bonds, which are financial instruments specifically designed to raise funds for environmentally sustainable projects. Green bonds have gained popularity as a way to attract investment in climate adaptation and mitigation projects.

By providing a clear link between investment and environmental outcomes, green bonds can help channel private capital into adaptation initiatives. Public-Private Partnerships (PPPs) also play a crucial role in financing adaptation projects. These partnerships can combine public sector oversight and funding with private sector expertise and efficiency, creating synergies that enhance the effectiveness and sustainability of adaptation efforts. For example, PPPs can be used to develop climate-resilient infrastructure, such as seawalls, stormwater management systems, and resilient transportation networks.

Role of International Aid and Cooperation

International aid and cooperation are vital for supporting adaptation efforts, particularly in developing countries. Developed nations have a responsibility to provide financial and technical assistance to vulnerable countries through international agreements and frameworks. For instance, the Paris Agreement includes commitments from developed countries to mobilize financial resources to assist developing countries in their adaptation and mitigation efforts. Bilateral and multilateral aid programmes can also support capacity-building initiatives, helping countries develop the institutional and technical capabilities needed to design and implement effective adaptation strategies. By fostering knowledge exchange and collaboration, international cooperation can enhance global resilience to climate change.

Integrating Adaptation into Development Planning

To effectively balance immediate economic needs with long-term adaptation goals, it is essential to integrate climate adaptation into broader development planning processes. This approach ensures that adaptation considerations are embedded in all aspects of economic and social development, rather than being treated as a separate or secondary issue.

For example, integrating climate risk assessments into infrastructure planning can help ensure that new developments are designed to withstand future climate impacts. Similarly, incorporating climate resilience into agricultural policies can promote the adoption of sustainable farming practices that enhance food security and reduce vulnerability to climate change.

Conclusion

Resource allocation for climate adaptation is a complex and multifaceted challenge, influenced by funding constraints and competing priorities. Securing adequate financial resources for adaptation projects is particularly difficult for developing countries, which often face additional economic and financial pressures. Balancing immediate economic needs with long-term adaptation goals requires innovative financing mechanisms, international cooperation, and the integration of adaptation into broader development planning. By addressing these challenges, governments and businesses can allocate resources more effectively, building resilience to climate change and ensuring sustainable economic growth.

Institutional Barriers

Institutional barriers pose significant challenges to effective economic adaptation to climate change. These barriers encompass regulatory hurdles and coordination issues, which can impede the development and implementation of adaptation strategies. Understanding and addressing these institutional barriers is crucial for creating a conducive environment for climate resilience.

Regulatory Hurdles

Existing policies and regulations can often hinder effective climate adaptation. Many current regulatory frameworks were developed without considering the long-term impacts of climate change, leading to misalignments that can obstruct adaptive measures. For instance, building codes and zoning laws may not account for future climate risks, resulting in infrastructure that is vulnerable to extreme weather events or sea-level rise. Moreover, regulatory environments can be slow to adapt to new information and changing conditions. The process of updating regulations to reflect the latest climate science can be cumbersome and politically contentious. This lag can leave communities and businesses exposed to climate risks that could have been mitigated with more timely regulatory adjustments. Another challenge is the inconsistency in regulations across different jurisdictions. Climate change impacts do not respect administrative boundaries, yet regulatory frameworks often vary significantly between regions. This inconsistency can complicate coordinated adaptation efforts, especially in areas where climate impacts cross regional or national borders. For example, a river basin spanning multiple country may be subject to differing water management policies, making it difficult to implement a cohesive strategy for managing climate-induced changes in water flow and quality. Additionally, some regulatory frameworks can create disincentives for adaptation. For example, subsidies for fossil fuels can discourage investment in renewable energy and other adaptation measures. Similarly, insurance regulations that do not account for increased climate risks can lead to under pricing of risk, reducing incentives for property owners to invest in resilience measures.

Coordination Issues

Effective climate adaptation requires coordinated action among a wide range of stakeholders, including government agencies, private sector entities, and civil society organizations. However, ensuring such coordination is often challenging due to differing priorities, objectives, and levels of resources among these groups. One major coordination issue is the fragmentation of responsibilities among various government agencies. Climate adaptation often falls under the purview of multiple departments, such as those responsible for environment, agriculture, infrastructure, and emergency management. Without clear mandates and mechanisms for inter-agency collaboration, efforts can become siloed, leading to inefficiencies and gaps in adaptation planning and implementation. The private sector also plays a critical role in climate adaptation, particularly through investments in resilient infrastructure, innovation in climate-smart technologies, and risk management practices.

However, aligning private sector actions with public adaptation goals requires effective communication and collaboration. Businesses may prioritize short-term financial performance over long-term resilience, and without incentives or regulatory requirements, they may underinvest in adaptation measures. Civil society organizations, including non-profits, community groups, and advocacy organizations, are essential for raising awareness, advocating for vulnerable populations, and implementing grassroots adaptation initiatives. However, these organizations often face resource constraints and may lack the influence needed to drive large-scale policy changes or secure significant funding for adaptation projects. To overcome coordination issues, it is essential to establish mechanisms for stakeholder engagement and collaboration.

This can include the creation of multi-stakeholder platforms that bring together representatives from government, business, and civil society to discuss and plan adaptation strategies. Such platforms can facilitate the sharing of information, resources, and best practices, enhancing the overall effectiveness of adaptation efforts. Furthermore, developing clear and consistent communication channels is vital for ensuring that all stakeholders are informed about climate risks and adaptation opportunities. This can involve public awareness campaigns, educational programmes, and the dissemination of climate data and research findings. By keeping stakeholders informed, it is possible to build a shared understanding of the challenges and the need for coordinated action.

Overcoming Institutional Barriers

Addressing institutional barriers to climate adaptation requires a multifaceted approach. Regulatory reforms are needed to update existing frameworks and remove disincentives for adaptive measures. This involves not only revising regulations to incorporate the latest climate science but also ensuring that policies are flexible enough to adapt to new information and changing conditions. Streamlining regulatory processes can also help expedite the implementation of adaptation projects. Enhancing coordination among stakeholders necessitates the development of robust institutional arrangements that facilitate collaboration.

This can include establishing inter-agency task forces, creating dedicated climate adaptation offices, and appointing climate adaptation coordinators to oversee and integrate efforts across different sectors and levels of government. Ensuring that these arrangements are adequately funded and supported is crucial for their success. International cooperation is also essential for addressing transboundary climate risks and sharing best practices. Multilateral agreements and frameworks can provide a platform for countries to coordinate their adaptation efforts, particularly in regions that share common climate challenges. International organizations and development agencies can play a key role in facilitating these collaborations and providing technical and financial support.

Conclusion

Institutional barriers, including regulatory hurdles and coordination issues, present significant challenges to effective climate adaptation. Overcoming these barriers requires regulatory reforms, enhanced stakeholder coordination, and robust institutional arrangements.

By addressing these institutional barriers, governments, businesses, and civil society can create a more conducive environment for climate resilience, ensuring that adaptation measures are effectively implemented and sustained over the long-term.

Social and Economic Inequalities

Climate change has profound implications for social and economic inequalities, often exacerbating existing disparities and creating new challenges for vulnerable populations. Addressing these inequalities through equitable adaptation strategies is essential to ensure that all communities can withstand and recover from climate impacts.

Vulnerability of Marginalized Groups

Climate change disproportionately affects marginalized and vulnerable populations, who often have the least capacity to adapt and recover from its impacts. These groups include low-income communities, indigenous peoples, women, children, the elderly, and people with disabilities. They face heightened risks due to a combination of socio-economic factors, geographic location, and limited access to resources and services. Low-income communities are particularly vulnerable because they often live in areas that are more susceptible to climate-related hazards, such as floodplains, coastal regions, and urban slums. These areas tend to have inadequate infrastructure, poor housing, and limited access to essential services like healthcare, clean water, and sanitation. As a result, when extreme weather events such as floods, hurricanes, or heat waves occur, these communities suffer disproportionately high levels of damage and loss.

For example, during Hurricane Katrina in 2005, low-income and predominantly African American neighbourhoods in New Orleans were among the hardest hit. The inadequate levee system, poor housing conditions, and lack of resources for evacuation and recovery exacerbated the disaster's impact on these communities. The long-term recovery process further highlighted the existing social and economic inequalities, as wealthier areas were able to rebuild more quickly and effectively. Indigenous peoples also face significant vulnerabilities to climate change due to their close dependence on natural resources for their livelihoods and cultural practices. Changes in climate patterns can disrupt traditional farming, fishing, and hunting practices, threatening food security and cultural heritage. Additionally, indigenous communities often reside in remote and environmentally sensitive areas, making them more susceptible to climate impacts such as sea-level rise, deforestation, and biodiversity loss.

Women and children are another group disproportionately affected by climate change. Women, particularly in developing countries, often bear the primary responsibility for household tasks such as water collection, food preparation, and care giving. Climate-related disruptions to water and food supplies can increase their workload and expose them to greater health risks. Children are also more vulnerable to the health impacts of climate change, including malnutrition, infectious diseases, and heat-related illnesses.

Equitable Adaptation

Ensuring that adaptation strategies are inclusive and equitable is a critical but complex challenge. Equitable adaptation requires recognizing and addressing the specific vulnerabilities and needs of different social groups, ensuring that adaptation measures do not inadvertently exacerbate inequalities. One approach to achieving equitable adaptation is through participatory planning and decision-making processes. Involving marginalized and vulnerable communities in the design and implementation of adaptation strategies can help ensure that their voices are heard and their needs are met. This participatory approach can also build local capacity and resilience by empowering communities to take ownership of adaptation initiatives. For instance, community-based adaptation projects have been successfully implemented in various regions, focusing on local knowledge and practices to enhance resilience. In Bangladesh, community-led initiatives have developed floating gardens and flood-resistant housing to adapt to frequent flooding.

These projects not only address immediate climate risks but also promote social cohesion and empower marginalized groups by involving them in decision-making processes. Equitable adaptation also requires targeted investments in infrastructure and services for vulnerable populations. This can include upgrading informal settlements with climate-resilient housing, improving access to clean water and sanitation, and developing early warning systems for extreme weather events. Ensuring that these investments reach the most vulnerable communities can help reduce their exposure to climate risks and enhance their adaptive capacity. Another key aspect of equitable adaptation is addressing the underlying social and economic determinants of vulnerability. This involves tackling issues such as poverty, education, healthcare, and access to resources, which can exacerbate the impacts of climate change on marginalized groups. For example, improving access to education and healthcare can enhance the overall resilience of communities by providing them with the knowledge and resources needed to adapt to changing conditions.

In addition, equitable adaptation strategies should consider gender-specific vulnerabilities and empower women as agents of change. This can involve promoting women's leadership in climate adaptation initiatives, ensuring their participation in decision-making processes, and providing targeted support for women's livelihoods. Studies have shown that women's involvement in natural resource management and disaster risk reduction can lead to more effective and sustainable outcomes.

Policy and International Support

Policymakers play a crucial role in promoting equitable adaptation by developing and implementing inclusive policies and programmes. This includes integrating equity considerations into national adaptation plans, setting up frameworks for monitoring and evaluating the impacts of adaptation measures on different social groups, and ensuring that adaptation funding is allocated to the most vulnerable communities. International support is also essential for promoting equitable adaptation, particularly in developing countries that may lack the financial and technical resources needed for effective adaptation. International climate finance mechanisms, such as the Green Climate Fund, can provide critical support for projects that target vulnerable populations and promote social equity. Additionally, international cooperation and knowledge-sharing can help disseminate best practices and innovative solutions for equitable adaptation.

Conclusion

Addressing social and economic inequalities is fundamental to ensuring effective and sustainable climate adaptation. Vulnerable and marginalized groups are disproportionately affected by climate change, and equitable adaptation strategies are needed to mitigate these impacts. This requires participatory planning, targeted investments, and addressing the underlying determinants of vulnerability. Policymakers and international support mechanisms play a vital role in promoting equity in adaptation efforts. By focusing on inclusivity and social justice, adaptation strategies cannot only enhance resilience but also contribute to broader goals of sustainable development and social equity.

Opportunities for Economic Growth and Innovation

The imperative to address climate change has spurred a wave of innovation and investment in green technologies and renewable energy.

These advancements present significant opportunities for economic growth and the creation of new industries, driving both environmental and economic benefits. This section explores the potential of green technologies and renewable energy in fostering economic development and job creation.

Innovation and Investment in Green Technologies

The transition to renewable energy sources and the development of green technologies are critical components of the global effort to mitigate climate change. This transition not only helps reduce greenhouse gas emissions but also stimulates economic growth by fostering innovation and attracting investment. Green technologies encompass a wide range of innovations designed to improve energy efficiency, reduce waste, and promote sustainable resource use. These include advancements in solar and wind energy, energy storage systems, electric vehicles, smart grids, and sustainable agriculture practices. Each of these technologies offers unique opportunities for innovation and economic expansion. For example, the solar energy sector has seen remarkable growth and innovation in recent years. Advances in photovoltaic technology have significantly increased the efficiency and affordability of solar panels, making solar energy a viable and competitive alternative to fossil fuels.

This progress has attracted substantial investment from both the private and public sectors. Companies are investing in research and development to create more efficient solar cells, while governments are providing subsidies and incentives to promote solar adoption. As a result, the global solar industry is experiencing rapid expansion, creating new markets and economic opportunities. Similarly, the development of energy storage technologies, such as advanced batteries, is crucial for integrating renewable energy into the grid and ensuring a stable energy supply. Innovations in battery technology are driving down costs and improving performance, enabling more widespread use of renewable energy. Companies like Tesla and Panasonic are at the forefront of this innovation, developing high-capacity batteries for both residential and commercial use. These advancements are attracting significant investment and driving economic growth in the energy storage sector.

Job Creation and Economic Development

Investment in sustainable industries can create new job opportunities and stimulate economic development. The shift towards renewable energy and green technologies requires a skilled workforce, leading to the creation of jobs across various sectors, including manufacturing, construction, research and development, and maintenance.

The renewable energy sector, in particular, has demonstrated significant potential for job creation. According to the International Renewable Energy Agency (IRENA), the renewable energy industry employed over 12 million people worldwide in 2020, a number that is expected to grow as the transition to clean energy accelerates. Jobs in renewable energy span a wide range of roles, from manufacturing solar panels and wind turbines to installing and maintaining renewable energy systems. For instance, the wind energy sector has become a major source of employment in many countries. In the United States, wind energy jobs have grown rapidly, with technicians and engineers in high demand to support the construction and maintenance of wind farms.

Similarly, in Europe, countries like Denmark and Germany have established themselves as leaders in wind energy technology, creating thousands of jobs and driving economic growth in the process. The electric vehicle (EV) industry is another area where green technology is creating substantial economic opportunities. The shift towards electric mobility is generating demand for new skills and jobs in battery manufacturing, EV production, and charging infrastructure development. Companies like Tesla, Nissan, and BYD are expanding their operations to meet the growing demand for electric vehicles, contributing to job creation and economic development in the automotive sector. In addition to direct job creation, investment in green technologies can have broader economic benefits through the development of supporting industries and supply chains.

For example, the growth of the renewable energy sector drives demand for raw materials, manufacturing equipment, and specialized services, creating economic opportunities across multiple sectors. This ripple effect can stimulate economic development in regions that invest in green technologies, leading to more resilient and diversified economies. Moreover, the transition to a green economy can enhance energy security and reduce dependence on imported fossil fuels. By investing in domestic renewable energy sources, countries can reduce their exposure to volatile global energy markets and improve their balance of trade. This shift can lead to more stable and sustainable economic growth, as well as increased national resilience to external economic shocks.

Case Studies: Economic Growth through Green Innovation

Several countries have successfully leveraged green technologies and renewable energy to drive economic growth and innovation.

Denmark, for example, has become a global leader in wind energy, with companies like Vestas and Ørsted at the forefront of wind turbine manufacturing and offshore wind farm development. Denmark's commitment to renewable energy has not only reduced its carbon footprint but also created thousands of jobs and positioned the country as an exporter of wind energy technology. China is another notable example, having rapidly expanded its renewable energy capacity and becomes the world's largest producer of solar panels and wind turbines. The Chinese government has made significant investments in renewable energy infrastructure, supporting the growth of domestic industries and creating millions of jobs.

This transition has helped China diversify its energy mix, reduce air pollution, and become a major player in the global renewable energy market. In the United States, states like California and Texas have embraced renewable energy as a driver of economic growth. California's ambitious clean energy policies have spurred innovation in solar and battery technologies, attracting investment and creating a thriving green economy. Texas, traditionally known for its oil and gas industry, has become a leader in wind energy, demonstrating how states can capitalize on their natural resources to drive economic development.

Conclusion

The transition to renewable energy and the development of green technologies present significant opportunities for economic growth and innovation. Investment in these sectors can create new job opportunities, stimulate economic development, and enhance energy security. Countries and regions that embrace green technologies and support sustainable industries are well-positioned to benefit from the economic and environmental advantages of the green economy. By fostering innovation and attracting investment, green technologies can drive a sustainable and prosperous future, addressing climate change while promoting economic growth.

Climate-Resilient Infrastructure

As the impacts of climate change become more pronounced, the development of climate-resilient infrastructure emerges as a crucial strategy for protecting economies and enhancing adaptive capacity. Building robust infrastructure that can withstand climate-induced stresses not only safeguards communities and economies but also opens up new avenues for business opportunities and innovation. This section explores the economic benefits of climate-resilient infrastructure development and the role of public-private partnerships in enhancing these projects.

Infrastructure Development

Climate-resilient infrastructure refers to the planning, design, and construction of physical structures that can endure the increasing frequency and intensity of extreme weather events, sea-level rise, and other climate-related challenges. Investing in such infrastructure is essential for mitigating the economic and social impacts of climate change, ensuring the continuity of services, and protecting lives and property. One significant area of focus is the improvement of transportation infrastructure, including roads, bridges, railways, and ports. These critical networks are vulnerable to climate impacts such as flooding, storms, and heatwaves. Upgrading transportation systems to withstand these stresses can prevent costly disruptions, facilitate efficient movement of goods and people, and maintain economic stability.

For example, elevating roadways in flood-prone areas, reinforcing bridges to withstand stronger winds, and designing rail systems to operate under extreme heat can enhance resilience. Water infrastructure is another critical area requiring investment. Climate change exacerbates the risk of both water scarcity and flooding, making the development of robust water management systems imperative. Investments in infrastructure such as dams, levees, and storm water drainage systems can protect communities from floods, while improved irrigation systems and water storage facilities can mitigate the effects of droughts.

The construction of desalination plants and the implementation of advanced water recycling technologies can also ensure a reliable supply of fresh water, even in regions facing severe water shortages. Urban infrastructure, particularly in rapidly growing cities, must be adapted to withstand climate impacts. This includes the development of green infrastructure, such as parks and green roofs, which can reduce urban heat island effects, manage storm water, and improve air quality. Building resilient housing and commercial buildings with energy-efficient designs, flood protection measures, and sustainable materials can also enhance urban resilience.

Economic Opportunities

Investing in climate-resilient infrastructure generates substantial economic opportunities. The construction and maintenance of resilient infrastructure create jobs across various sectors, including engineering, construction, and manufacturing. These projects require skilled labour, driving demand for training and education programmes that equip workers with the necessary expertise. Furthermore, climate-resilient infrastructure projects can stimulate innovation in materials science, engineering, and technology.

The need for durable and sustainable building materials has led to the development of advanced composites, high-performance concrete, and other innovative solutions. Similarly, the integration of smart technologies into infrastructure, such as sensors for real-time monitoring and adaptive control systems, enhances resilience and efficiency. The private sector can also benefit from opportunities related to climate-resilient infrastructure. Companies specializing in construction, engineering, and technology can expand their markets by developing and deploying solutions tailored to climate resilience.

Financial institutions can create new investment products and services that support resilient infrastructure projects, attracting capital from investors seeking sustainable and profitable ventures. Moreover, resilient infrastructure can enhance the overall attractiveness of regions and cities, encouraging economic growth and development. Businesses are more likely to invest in areas with reliable infrastructure, as it reduces operational risks and enhances supply chain stability. This can lead to increased economic activity, higher property values, and greater competitiveness on a global scale.

Public-Private Partnerships

Public-Private Partnerships (PPPs) play a crucial role in the development and implementation of climate-resilient infrastructure projects. These collaborations leverage the strengths and resources of both sectors to achieve more effective and efficient outcomes. Governments often face budget constraints and limited technical capacity, making it challenging to finance and manage large-scale infrastructure projects independently. By partnering with the private sector, governments can access additional funding, expertise, and innovative solutions. Private companies, in turn, benefit from the stability and scale provided by public projects, as well as opportunities for long-term returns on investment.

One successful example of a PPP in climate-resilient infrastructure is the Thames Estuary 2100 project in the United Kingdom. This initiative aims to protect London and the surrounding areas from flooding due to sea-level rise and storm surges. The project involves collaboration between the UK government, local authorities, and private sector partners, combining public funding with private investment and expertise. The result is a comprehensive and adaptive flood management system that includes barriers, embankments, and sustainable urban drainage solutions. Another example is the Dutch Room for the River programme, which focuses on enhancing flood resilience in the Netherlands.

This programme, involving public authorities and private engineering firms, has implemented measures such as creating floodplains, relocating dikes, and constructing overflow channels. These interventions not only protect communities from flooding but also create new recreational and ecological areas, demonstrating the multifunctional benefits of climate-resilient infrastructure. PPPs can also facilitate knowledge transfer and capacity building. By working together, public and private entities can share best practices, technical expertise, and innovative approaches to infrastructure development. This collaboration enhances the overall quality and effectiveness of resilience projects, ensuring that they are designed and implemented to the highest standards.

Conclusion

Building climate-resilient infrastructure is essential for protecting economies from the adverse impacts of climate change while creating significant economic opportunities. Investments in resilient transportation, water, and urban infrastructure can safeguard communities, stimulate job creation, and drive innovation. Public-private partnerships are vital for enhancing the effectiveness of these projects, leveraging the strengths and resources of both sectors to achieve sustainable and resilient outcomes. By prioritizing climate-resilient infrastructure, societies can enhance their adaptive capacity, promote economic growth, and ensure a sustainable future.

Sustainable Agriculture

The agricultural sector faces significant challenges due to climate change, including shifting weather patterns, increased frequency of extreme weather events, and changing pest and disease dynamics. However, these challenges also present opportunities for innovation and economic diversification. By adopting sustainable farming practices and technologies, the agricultural sector can enhance food security, boost productivity, and reduce vulnerability to climate impacts.

Agricultural Innovation

Advancements in sustainable farming practices and technologies are essential for building a resilient agricultural sector. Innovations such as precision agriculture, agro-ecological methods, and climate-smart farming can significantly improve productivity and sustainability.

- **Precision Agriculture:** This approach involves the use of technology, such as GPS, drones, and IoT sensors, to monitor and manage crops with high precision.
  Precision agriculture enables farmers to optimize inputs like water, fertilizers, and pesticides, leading to higher yields and reduced environmental impact. For example, using drones to monitor crop health and detect pest infestations early can prevent losses and reduce the need for chemical interventions.
- **Agro-ecological Methods:** These practices focus on integrating ecological principles into farming systems. Techniques such as crop rotation, intercropping, and the use of cover crops can enhance soil health, increase biodiversity, and improve resilience to climate variability. Agro-ecology promotes sustainable resource use and reduces dependency on external inputs, making farming systems more robust and adaptable.

- **Climate-Smart Agriculture:** This approach involves adopting practices that enhance resilience to climate change while mitigating greenhouse gas emissions. Techniques such as conservation tillage, integrated pest management, and the use of drought-resistant crop varieties can help farmers adapt to changing conditions. For instance, the development and adoption of heat-tolerant and drought-resistant crops can ensure stable yields even under adverse weather conditions.

Economic Diversification

Diversifying agricultural economies to include climate-resilient crops and practices is crucial for reducing vulnerability to climate impacts. By broadening the range of crops and farming activities, farmers can spread risk and enhance their resilience.

- **Climate-Resilient Crop:** Introducing crops that are more tolerant to climate stresses, such as drought, heat, and salinity, can help farmers maintain productivity in the face of climate change. For example, cultivating millet and sorghum, which are more drought-tolerant than traditional staples like wheat and rice, can improve food security in arid regions.
- **Agro-forestry:** This practice involves integrating trees and shrubs into agricultural landscapes. Agro-forestry systems can provide multiple benefits, including enhanced biodiversity, improved soil fertility, and increased carbon sequestration. Trees can also act as windbreaks and provide shade, protecting crops from extreme weather conditions.
- **Value-added Products:** Farmers can diversify their income streams by producing value-added products, such as processed foods, organic produce, and bio-based materials. This approach not only increases profitability but also reduces dependency on single crops, enhancing economic resilience.

Insurance and Financial Services

The financial sector plays a pivotal role in managing climate risks and funding adaptation projects. Developing innovative insurance products and financial services tailored to climate risks can enhance economic resilience and provide security for businesses and communities.

Risk Management

Innovative insurance products designed to address climate risks can help protect farmers, businesses, and communities from the financial impacts of extreme weather events and other climate-related challenges.

- **Index-based Insurance:** This type of insurance pays out benefits based on predefined indices, such as rainfall levels or temperature thresholds, rather than actual loss assessments. Index-based insurance is particularly useful for small-holder farmers, as it reduces the administrative burden and provides quicker payouts. For example, a drought index insurance policy could provide payouts when rainfall falls below a certain threshold, helping farmers cope with crop losses.
- **Catastrophe Bonds:** These are financial instruments that provide payouts in the event of a catastrophic event, such as a hurricane or earthquake. Catastrophe bonds transfer risk from insurers to investors, spreading the financial impact of disasters. They can be an effective tool for managing large-scale climate risks and ensuring rapid recovery after extreme events.

Investment Opportunities

Financial markets can play a crucial role in funding climate adaptation projects through instruments such as green bonds and climate funds.

- **Green Bonds:** These bonds are specifically designed to fund projects that have positive environmental benefits. Green bonds can be used to finance a wide range of climate adaptation and mitigation projects, including renewable energy installations, energy efficiency improvements, and resilient infrastructure developments. The market for green bonds has grown significantly in recent years, reflecting increasing investor interest in sustainable finance.
- **Climate Funds:** These funds pool resources from various sources, including governments, international organizations, and private investors, to finance climate adaptation and mitigation projects. Climate funds can provide grants, loans, and technical assistance to support projects that enhance resilience and reduce greenhouse gas emissions. Examples include the Green Climate Fund (GCF) and the Adaptation Fund, which provide financial support for climate projects in developing countries.
- **Sustainable Investing:** There is a growing trend among investors to prioritize environmental, social, and governance (ESG) factors in their investment decisions. Sustainable investing focuses on companies and projects that demonstrate strong ESG performance, including those that contribute to climate resilience. This approach not only supports climate adaptation efforts but also aligns with the increasing demand for responsible and ethical investment options.

Case Studies: Financial Innovation in Climate Adaptation

Several initiatives worldwide show the potential of insurance and financial services in enhancing climate resilience.

- **African Risk Capacity (ARC):** This specialized agency of the African Union provides parametric insurance to African countries to help them cope with climate risks such as droughts and floods. ARC uses weather data to trigger payouts, ensuring that funds are available quickly when a disaster strikes. This mechanism helps governments respond more effectively to emergencies and reduces the economic impact of climate shocks.

- **Forest Resilience Bond (FRB):** This innovative financial instrument developed by Blue Forest Conservation is designed to finance forest restoration projects that reduce the risk of wildfires. The FRB pools capital from investors to fund forest management activities, such as thinning and controlled burns. These activities reduce the likelihood of catastrophic wildfires, protecting communities and ecosystems while providing returns to investors.
- **Insurance Resilience Global Partnership:** This initiative aims to strengthen the resilience of developing countries by expanding access to climate and disaster risk insurance. The partnership brings together governments, international organizations, and the private sector to develop and implement insurance solutions that protect vulnerable populations from climate risks.

Conclusion

The development of sustainable agriculture and innovative financial services present significant opportunities for enhancing climate resilience and economic growth. By adopting advanced farming practices and technologies, the agricultural sector can boost productivity and reduce vulnerability to climate impacts. Similarly, innovative insurance products and financial instruments can provide crucial risk management and funding for adaptation projects. Embracing these opportunities is essential for building a resilient and sustainable future in the face of climate change.

Case Studies on Economic Adaptation Strategies

The Netherlands: Flood Management and Infrastructure

The Netherlands, a country largely situated below sea level, has long been at the forefront of flood management and infrastructure development. Its innovative solutions and strategic investments have not only protected the nation from the adverse impacts of climate change but also stimulated economic growth and positioned the Netherlands as a global leader in water management.

Innovative Solutions

The Netherlands has developed and implemented a range of advanced flood management systems designed to protect its low-lying areas from sea-level rise, storm surges, and extreme weather events. Two notable examples of these innovations are the Delta Works and adaptive delta management.

- **Delta Works**: This extensive system of dams, sluices, locks, dykes, and storm surge barriers is considered one of the most sophisticated flood protection systems in the world. Completed over several decades following the catastrophic North Sea flood of 1953, the Delta Works were designed to protect the south-western part of the Netherlands. Key structures include the Maeslantkering and the Oosterscheldekering. The Maeslantkering is a movable storm surge barrier that can close off the New Waterway, protecting Rotterdam and surrounding areas from high water levels. The Oosterscheldekering, another massive storm surge barrier, is capable of controlling water flow while maintaining tidal movements crucial for the local ecosystem.
- **Adaptive Delta Management:** Recognizing the need for flexibility in the face of uncertain future climate conditions, the Netherlands has adopted adaptive delta management principles. This approach involves dynamic planning and the use of scenarios to anticipate future changes. The "Room for the River" programme, for instance, aims to give rivers more space to accommodate higher water levels. This involves measures such as relocating dykes, creating overflow areas, and deepening riverbeds. By allowing rivers to expand naturally during periods of high water,

this strategy reduces the risk of flooding and enhances ecological resilience.

Economic Impact

The economic impact of the Netherlands' flood management strategies extends beyond mere protection against natural disasters. These measures have catalyzed economic growth, driven innovation, and solidified the country's reputation as a leader in water management.

- **Stimulating Innovation:** The development and maintenance of advanced flood management systems have spurred innovation in engineering, materials science, and environmental management. Dutch companies and research institutions have become pioneers in these fields, creating new technologies and methodologies that can be applied globally. The expertise gained through domestic projects has enabled Dutch firms to export their knowledge and services worldwide, contributing to economic growth.
- **International Expertise:** Dutch water management expertise is in high demand internationally. Dutch companies and consultancies are involved in flood protection projects in countries such as the United States, Vietnam, and Bangladesh. For instance, Dutch experts have contributed to the design of the New Orleans flood protection system post-Hurricane Katrina and the development of the Mekong Delta Plan in Vietnam. These international projects generate significant revenue and enhance the global influence of Dutch engineering and environmental consultancy firms.
- **Economic Resilience:** The investments in flood management infrastructure have also increased the overall economic resilience of the Netherlands. By safeguarding critical infrastructure, agricultural lands, and urban areas, these measures prevent costly damage and ensure the continuity of economic activities. The ability to

maintain operational ports, industrial areas, and residential zones during extreme weather events is crucial for the Netherlands' economic stability and growth.

- **Tourism and Recreation:** Innovative flood management projects have also created new opportunities for tourism and recreation. The Delta Works, for example, attract tourists interested in engineering marvels and sustainable infrastructure. Furthermore, the creation of natural floodplains and recreational areas as part of adaptive delta management initiatives enhances the quality of life for residents and attracts visitors, contributing to local economies.

Conclusion

The Netherlands' approach to flood management and infrastructure development offers a compelling case study in economic adaptation to climate change. Through the implementation of innovative solutions like the Delta Works and adaptive delta management, the country has effectively protected itself from the adverse impacts of rising sea levels and extreme weather. These measures have not only safeguarded infrastructure but also stimulated economic growth through innovation, international expertise, and increased economic resilience. The success of the Netherlands in managing flood risks underscores the potential for other nations to develop and implement similar strategies, leveraging their unique geographic and economic contexts to build resilience and drive economic development.

Bangladesh: Community-Based Adaptation

Bangladesh, one of the most climate-vulnerable countries in the world, has developed innovative community-based adaptation strategies to address the challenges posed by flooding and sea-level rise. These grassroots initiatives not only enhance resilience to climate impacts but also offer significant social and economic benefits, improving the livelihoods and well-being of local populations.

Grassroots Initiatives

Bangladesh's community-based adaptation strategies focus on leveraging local knowledge and resources to develop sustainable solutions. Two prominent examples are the construction of floating gardens and the implementation of raised housing.

- **Floating Gardens:** Floating gardens, or "baira", are an ingenious agricultural technique used in the flood-prone regions of Bangladesh. These gardens are constructed using water hyacinths and other buoyant plant materials to create a floating platform. On these platforms, farmers grow vegetables and other crops, even during the monsoon season when much of the land is submerged. This method not only ensures continuous food production despite flooding but also utilizes abundant local resources. The floating gardens are particularly beneficial in regions like the Haor basin, where seasonal flooding is severe. This technique has been practised for centuries and has been revitalized and supported by various non-governmental organizations (NGOs) and government programmes. The gardens provide a reliable source of food and income for families, enhancing their food security and economic resilience.
- **Raised Housing:** Another critical adaptation strategy in Bangladesh is the construction of raised housing, known locally as "plinths". In flood-prone areas, homes are built on elevated platforms to protect them from floodwaters. This approach not only prevents property damage but also ensures that families can remain in their homes during flood events, reducing the need for displacement and temporary shelters. Raised housing projects often involve community participation and support from NGOs and government agencies. For example, the Comprehensive Disaster Management Programme (CDMP) has facilitated the construction of thousands of raised homes in vulnerable

regions. These elevated structures are often accompanied by community infrastructure improvements, such as raised roads and flood shelters, further enhancing the resilience of the communities.

Social and Economic Benefits

The community-based adaptation strategies employed in Bangladesh offer a range of social and economic benefits, contributing to the overall resilience and development of local populations. Improved Food Security Floating gardens significantly enhance food security in flood-prone areas. By ensuring year-round agricultural production, these gardens reduce the risk of food shortages during the monsoon season. The cultivation of diverse crops on floating gardens also improves dietary diversity and nutritional intake for families. Additionally, the surplus produce from these gardens can be sold in local markets, generating income for households. Enhanced Community Resilience Community-based adaptation strategies foster resilience by empowering local populations to manage and mitigate climate risks.

The involvement of community members in the planning and implementation of adaptation projects ensures that solutions are tailored to local needs and conditions. This participatory approach enhances the effectiveness and sustainability of adaptation measures. Raised housing projects, for example, protect families from the immediate impacts of flooding, allowing them to continue their daily activities and maintain their livelihoods. The stability provided by these elevated homes reduces the physical and psychological stress associated with displacement and property loss. Furthermore, community infrastructure improvements, such as raised roads and flood shelters, ensure that essential services and emergency assistance are accessible during flood events. Economic Opportunities Community-based adaptation initiatives create new economic opportunities for local populations. The construction and maintenance of floating gardens and raised housing provide employment and income-generating activities.

Local artisans and labourers are often involved in building raised homes and community infrastructure, contributing to the local economy. Additionally, the agricultural productivity of floating gardens opens up new market opportunities for farmers. By producing crops that can be sold locally or transported to larger markets, farmers can diversify their income sources and reduce their economic vulnerability. The skills and knowledge gained through these adaptation projects can also be transferred to other areas, promoting broader economic development. Social Cohesion and Empowerment Community-based adaptation strategies strengthen social cohesion and empower local populations.

The collaborative nature of these projects fosters a sense of community and shared responsibility. By working together to address common challenges, community members build strong social networks and support systems. The involvement of women in adaptation projects is particularly noteworthy. In many cases, women play a central role in the management of floating gardens and the construction of raised housing. This participation enhances gender equity and empowers women to take on leadership roles within their communities. The recognition of women's contribution to adaptation efforts also promotes social inclusion and equity.

Conclusion

Bangladesh's community-based adaptation strategies offer a powerful example of how grassroots initiatives can effectively address the challenges posed by climate change. By leveraging local knowledge and resources, Bangladesh has developed innovative solutions like floating gardens and raised housing to enhance resilience to flooding and sea-level rise. These strategies provide significant social and economic benefits, including improved food security, enhanced community resilience, economic opportunities, and social cohesion. The success of these community-based approaches highlights the importance of involving local populations in the planning and implementation of adaptation projects. By empowering communities to take an active role in managing climate risks, Bangladesh has not only improved the well-being of its citizens but also set a precedent for other countries facing similar challenges. As climate change continues to impact vulnerable regions worldwide, the lessons learned from Bangladesh's experience will be invaluable in shaping effective and sustainable adaptation strategies globally.

California: Renewable Energy Transition

California has emerged as a global leader in the renewable energy transition through a combination of robust policy leadership and substantial investment in clean energy infrastructure and technology. These efforts have not only significantly reduced greenhouse gas emissions but also transformed the state's economy, creating thousands of green jobs and positioning California at the forefront of the global clean energy market.

Policy Leadership

California's commitment to renewable energy is evident in its aggressive and forward-thinking policies designed to foster the growth of the clean energy sector. The state has implemented a series of legislative measures and regulatory frameworks aimed at reducing carbon emissions, promoting energy efficiency, and encouraging the adoption of renewable energy sources.

- **Renewable Portfolio Standard (RPS):** One of the cornerstone policies is California's Renewable Portfolio Standard, which mandates that a certain percentage of the state's electricity must come from renewable sources. Initially set at 20 per cent by 2010, the target has been progressively increased over the years. In 2018, the state passed Senate Bill 100, which set an ambitious goal of achieving 100 per cent clean, carbon-free electricity by 2045. This policy has driven significant investment in renewable energy projects, including solar, wind, geothermal, and biomass energy.

- **Cap-and-Trade Programme:** California's cap-and-trade programme, established under the Global Warming Solutions Act of 2006 (AB 32), is another critical component of the state's climate strategy. The programme sets a state-wide limit on greenhouse gas emissions and allows businesses to buy and sell emission allowances. This market-based approach provides economic incentives for companies to reduce their carbon footprint and invest in cleaner technologies.
- **Incentives and Subsidies:** The state has also implemented various incentives and subsidies to encourage the adoption of renewable energy. Programmes such as the California Solar Initiative and the Self-Generation Incentive Programme offer financial incentives for installing solar panels and other renewable energy systems. Additionally, tax credits and rebates are available for energy-efficient appliances, electric vehicles, and energy storage solutions.

Economic Transformation

California's investment in renewable energy infrastructure and technology has had a profound impact on the state's economy, driving economic growth, job creation, and technological innovation.

- **Job Creation:** The renewable energy sector has become a major source of employment in California. According to the California Energy Commission, the state's clean energy industry employs over half a million workers. Solar energy alone accounts for a significant portion of these jobs, with California leading the nation in solar installations. The construction, maintenance, and operation of renewable energy projects provide employment opportunities across various skill levels, from engineers and technicians to construction workers and administrative staff.
- **Economic Growth:** Investment in renewable energy has stimulated economic growth by attracting billions

of dollars in capital investment. California's supportive policies and favourable market conditions have made it an attractive destination for clean energy companies and investors. The state's clean energy sector has seen substantial growth in venture capital funding, particularly in areas such as solar technology, energy storage, and electric vehicles. This influx of capital has spurred innovation, leading to the development of cutting-edge technologies and the establishment of numerous start-ups and research institutions.

- **Technological Innovation:** California's focus on renewable energy has positioned it as a hub for technological innovation in the clean energy sector. The state is home to leading research institutions, such as the California Institute of Technology (Caltech), Stanford University, and the University of California system, which conduct pioneering research in renewable energy and sustainability. Collaborations between academia, industry, and government have accelerated the development and deployment of new technologies, such as advanced solar panels, smart grids, and energy storage systems.
- **Global Leadership:** California's leadership in renewable energy has extended beyond its borders, influencing global clean energy markets and policies. The state's ambitious goals and successful implementation strategies have served as a model for other regions and countries. California has established partnerships and agreements with other states and nations to promote renewable energy and share best practices. For instance, the Under2 Coalition, co-founded by California, brings together sub-national governments committed to reducing greenhouse gas emissions in line with the Paris Agreement.
- **Energy Independence and Security:** The transition to renewable energy has also enhanced California's energy independence and security. By reducing reliance on

fossil fuels and diversifying its energy sources, the state has become less vulnerable to price volatility and supply disruptions in global energy markets. The growth of distributed energy resources, such as rooftop solar panels and community micro grids, has further increased the resilience of California's energy system.

Conclusion

California's renewable energy transition exemplifies the transformative potential of proactive climate policies and strategic investments in clean energy infrastructure. Through a combination of policy leadership, economic incentives, and technological innovation, the state has successfully reduced greenhouse gas emissions, created thousands of green jobs, and stimulated economic growth. California's experience demonstrates that ambitious climate action can drive economic transformation, positioning regions as leaders in the global clean energy market. As other states and countries seek to transition to sustainable energy systems, California's model offers valuable insights into the pathways and benefits of embracing renewable energy.

Rwanda: Climate-Resilient Agriculture

Rwanda, a landlocked country in East Africa, has faced significant challenges due to climate change, including unpredictable rainfall patterns, soil erosion, and land degradation. In response, the country has implemented comprehensive agricultural reforms focused on climate-smart practices. These initiatives aim to enhance resilience to climate change, improve food security, and promote sustainable land management. The results have been both economically and environmentally transformative, demonstrating the potential of climate-resilient agriculture.

Agricultural Reforms

Rwanda's approach to climate-resilient agriculture includes a variety of innovative practices designed to adapt to changing climatic conditions and mitigate environmental degradation. Key strategies include agro-forestry, soil conservation techniques, and the promotion of sustainable farming practices.

- **Agro-forestry:** Agro-forestry, the integration of trees and shrubs into agricultural landscapes, is a cornerstone of Rwanda's climate-smart agriculture initiatives. This practice offers multiple benefits, such as improving soil fertility, reducing erosion, and providing additional sources of income through the production of fruits, nuts, and timber. By planting trees alongside crops, farmers can create more resilient agricultural systems that are better equipped to withstand climate variability. The Rwandan government, in collaboration with various international organizations, has supported large-scale agro-forestry projects, providing training and resources to local farmers.
- **Soil Conservation Techniques:** Soil erosion and degradation have been major challenges for Rwandan agriculture. To address these issues, the government has promoted the adoption of soil conservation techniques, such as terracing, contour farming, and the use of cover crops. Terracing involves creating stepped levels on slopes to reduce runoff and soil erosion, while contour farming aligns planting along the natural contours of the land to minimize erosion. Cover crops, planted during off-seasons, protect the soil from erosion and enhance its fertility through nitrogen fixation. These practices help maintain soil health, increase water retention, and improve crop yields.
- **Sustainable Farming Practices:** Rwanda has also emphasized the importance of sustainable farming practices to enhance agricultural resilience. This includes the use of organic fertilizers, integrated pest management, and crop diversification. Organic fertilizers, such as compost and manure, improve soil fertility without the environmental drawbacks associated with chemical fertilizers. Integrated

pest management combines biological, cultural, and mechanical methods to control pests, reducing the need for chemical pesticides. Crop diversification, the practice of growing a variety of crops, reduces the risk of total crop failure due to pests or adverse weather conditions and improves food security.

Economic and Environmental Gains

The implementation of climate-smart agricultural practices in Rwanda has yielded significant economic and environmental benefits, contributing to the country's overall resilience and sustainable development.

- **Increased Agricultural Productivity:** One of the most notable economic benefits of Rwanda's climate-resilient agriculture initiatives is the increase in agricultural productivity. The adoption of agro-forestry, soil conservation techniques, and sustainable farming practices has led to higher crop yields and more reliable harvests. For example, farmers who have implemented terracing and organic fertilization have reported substantial improvements in crop output. This increase in productivity has not only boosted farmers' incomes but also enhanced the country's food security.
- **Improved Food Security:** Climate-smart agriculture has played a crucial role in improving food security in Rwanda. By increasing the resilience of agricultural systems to climate variability, these practices ensure more consistent food production. The integration of diverse crops and agro-forestry products has also improved the nutritional quality of the food available to local communities. As a result, households are less vulnerable to food shortages and price fluctuations caused by climate-related disruptions.
- **Sustainable Land Management:** The environmental gains from Rwanda's agricultural reforms are equally

impressive. Agro-forestry and soil conservation techniques have significantly reduced soil erosion and degradation, preserving the fertility and productivity of agricultural land. The planting of trees and use of cover crops enhance biodiversity and promote healthier ecosystems. Additionally, the reduction in chemical fertilizer and pesticide use minimizes environmental pollution, protecting water resources and wildlife. These sustainable land management practices contribute to the long-term viability of Rwanda's agricultural sector and the overall health of its environment.

- **Economic Diversification:** The promotion of climate-smart agriculture has also facilitated economic diversification in rural areas. By introducing farmers to new crops and products, such as fruits, nuts, and timber from agro-forestry systems, the reforms have created additional sources of income. This diversification reduces farmers' dependence on a single crop, making them less vulnerable to climate-related risks and market fluctuations. Furthermore, the skills and knowledge gained through training and capacity-building initiatives enhance farmers' ability to innovate and adapt to changing conditions.
- **Community Empowerment and Gender Inclusion:** Rwanda's climate-resilient agriculture initiatives have had positive social impacts, particularly in terms of community empowerment and gender inclusion. Many of the projects involve community-based approaches, encouraging collective action and local decision-making. Women, who play a crucial role in agriculture, have been actively involved in these initiatives, gaining access to training, resources, and leadership opportunities. This empowerment has not only improved the economic status of women but also strengthened the social fabric of rural communities.

Conclusion

Rwanda's commitment to climate-resilient agriculture exemplifies how comprehensive and well-implemented reforms can address the challenges posed by climate change while delivering substantial economic and environmental benefits. Through the adoption of agro-forestry, soil conservation techniques, and sustainable farming practices, Rwanda has enhanced agricultural productivity, improved food security, and promoted sustainable land management. These initiatives have not only made the country's agricultural sector more resilient to climate variability but also contributed to economic diversification and community empowerment. The success of Rwanda's climate-smart agriculture strategies provides valuable lessons for other countries facing similar challenges. By prioritizing sustainable practices and investing in the capacity of local communities, nations can build more resilient agricultural systems that support both economic development and environmental sustainability. Rwanda's experience underscores the potential of climate-resilient agriculture to drive positive change and ensure a sustainable future for vulnerable regions.

REFERENCES

Auerbach, A.J., and Gorodnichenko, Y. (2022). "Fiscal Multipliers in Recessions and Expansions: Does It Matter Whether Government Spending is Increasing or Decreasing?" *Journal of Economic Perspectives, 36*(1): 41-64.

Ball, L.M., Gagnon, E., Honohan, P. and Krogstrup, S., (2022). "What Else Can Central Banks Do?" *IMF Economic Review, 70*(1): 167-206.

Bianchi, F. and Melosi, L. (2022). "Inflation Expectations and the Supply Chain". *American Economic Review, 112*(4): 1284-1320.

Blanchard, O. (2023). "The Policy Toolkit for Inflation: Interest Rates, Fiscal Policy, and Supply Side Measures". *Journal of Policy Modelling, 45*(2): 203-217.

Christiano, L.J., Eichenbaum, M.S. and Trabandt, M. (2022). "Understanding the Great Inflation: Lessons for Current Policy". *Quarterly Journal of Economics, 137*(3): 1263-1300.

Clarida, R.H. (2022). "Monetary Policy, Price Stability, and Inflation: Theoretical Perspectives and Practical Lessons". *Journal of Economic Literature, 60*(1): 34-62.

Coibion, O., Gorodnichenko, Y. and Weber, M. (2022). "How Inflation Expectations Affect Households and Firms: The Role of Central Bank Communication". *Brookings Papers on Economic Activity, 2022*(1): 115-174.

De Long, J.B. and Summers, L.H. (2022). "The Macroeconomic Policy Response to the COVID-19 Pandemic: The Case for a Higher Inflation Target". *Economic Policy, 37*(111): 409-451.

Guvenen, F., Kaplan, G. and Song, J. (2022). "How Does Inflation Affect Different Income Groups?" *Journal of Political Economy, 130*(5): 1437-1473.

Hamilton, J.D., (2022). "Why You Should Never Use the Hodrick-Prescott Filter". *Review of Economic Studies, 89*(1): 104-120.

Jordà, Ò. and Taylor, A.M. (2022). "The Pandemic, Inflation, and Debt Sustainability". *Economic Policy, 37*(110): 275-314.

Laubach, T. and Williams, J.C. (2022). "Measuring the Natural Rate of Interest: International Trends and Determinants". *Brookings Papers on Economic Activity, 2022*(3): 57-108.

Mishkin, F.S. (2022). "Monetary Policy Strategy: Lessons from the Financial Crisis". *Journal of Economic Perspectives, 36*(3): 123-148.

Mussa, M. (2022). "Inflation, Real Interest Rates, and the Real Exchange Rate: An Empirical Investigation". *Journal of Monetary Economics, 127*: 243-271.

Reis, R. (2022). "The Analytics of Monetary Policy: A New Keynesian Perspective". *Quarterly Journal of Economics, 137*(4): 1951-1987.

2

Historical Perspectives on Economic Adaptation to Climate Change

Early Examples of Economic Adaptation to Climate Variations

Economic adaptation to climate variations is not a modern phenomenon; it has been an essential aspect of human survival and development throughout history. Early societies had to cope with changing climatic conditions long before the advent of advanced technology and scientific understanding. These early examples provide valuable insights into how human communities have historically managed climate-induced challenges.

Adaptation in Ancient Civilizations

Ancient civilizations across different regions of the world developed unique methods to adapt to the changing climatic conditions, ensuring their survival and economic stability. For instance, the ancient Mesopotamians, Egyptians, and Indus Valley inhabitants, living in regions prone to flooding and droughts, created sophisticated water management systems to secure their agricultural production.

Mesopotamia

In Mesopotamia, the land between the Tigris and Euphrates rivers, the unpredictable and often severe flooding was a significant challenge. The Mesopotamians constructed intricate irrigation canals, dikes, and reservoirs to control water flow and manage their water resources efficiently. These systems allowed them to regulate flooding, store water for dry periods, and irrigate their crops, ensuring a stable food supply and fostering economic growth. Their ability to adapt to the region's climate variability through engineering and communal labour exemplifies early economic adaptation.

Egypt

Similarly, in ancient Egypt, the Nile River's annual flooding was both a blessing and a challenge. The Egyptians developed a calendar system based on the Nile's flood cycles, which helped them plan agricultural activities effectively.

They also built extensive irrigation networks to channel floodwaters into their fields, maximizing agricultural output and supporting a growing population. The predictability of the Nile's flooding, coupled with the Egyptians' adaptive strategies, allowed them to flourish economically.

Indus Valley

In the Indus Valley, the ancient civilization of Harappa and Mohenjo-Daro faced similar challenges. They developed advanced urban planning and water management techniques, including sophisticated drainage systems and reservoirs. These innovations enabled them to control water supply, manage floods, and ensure a stable agricultural base. Their ability to adapt to the climatic conditions contributed to the civilization's economic prosperity.

Nomadic Adaptations

In regions where sedentary agriculture was less viable due to harsh climatic conditions, such as arid and semi-arid zones, nomadic pastoralism emerged as a successful adaptation strategy. Nomadic communities, particularly in the Arabian Peninsula and Central Asia, adapted to the climate by relying on the mobility of their livestock.

- **Arabian Peninsula:** The Bedouin tribes of the Arabian Peninsula exemplify how nomadic pastoralism can serve as an economic adaptation to climate variability. These tribes moved seasonally with their herds of camels, sheep, and goats to find water and grazing land. Their mobility allowed them to exploit different ecological niches and avoid the harshest conditions, ensuring the survival of both the people and their livestock. This lifestyle required deep knowledge of the environment, including the locations of water sources and the timing of seasonal changes, demonstrating a sophisticated understanding of their climate and resources.
- **Central Asia:** Similarly, in Central Asia, the nomadic tribes adapted to the region's harsh climatic conditions by developing a migratory lifestyle. The Mongols, for instance, moved across vast steppes with their herds, following the seasonal availability of pasture and water. This mobility reduced pressure on any single area, preventing overgrazing and allowing the land to recover. The economic adaptation of these nomadic societies ensured their survival and allowed them to thrive in environments where sedentary agriculture would have been unsustainable.

Agricultural Innovations

In regions with challenging terrain, such as mountainous areas, early societies developed agricultural innovations to adapt to the environment. Terrace farming is one such examples, used in various parts of the world, including the Andean highlands of South America and the Himalayan region.

- **Andean Highlands:** The Incas of the Andean highlands mastered the technique of terrace farming, creating stepped agricultural fields on steep mountain slopes. These terraces reduced soil erosion and improved water retention, making agriculture possible in otherwise inhospitable terrain. The terraces also created microclimates that protected crops from extreme weather conditions. This adaptation not only supported a stable food supply but also allowed the Incas to develop a complex and prosperous society.
- **Himalayan Region:** In the Himalayan region, terrace farming has been practised for centuries. The terraced fields conserve soil and water, allowing farmers to cultivate crops such as rice, maize, and millet on steep slopes. This method of farming maximizes the use of available land and reduces the risk of landslides. By adapting their agricultural practices to the mountainous environment, these communities have maintained sustainable livelihoods for generations.

Lessons from Early Adaptation Strategies

The historical examples of economic adaptation to climate variations provide valuable lessons for contemporary efforts to address climate change. These early strategies highlight the importance of innovation, community involvement, and sustainability in developing effective adaptation measures.

- **Innovation and Technology:** The ability of early societies to innovate and develop new technologies was crucial for their adaptation success. The advanced water management systems of Mesopotamia and Egypt, the sophisticated urban planning of the Indus Valley, and the agricultural innovations of the Incas all demonstrate how technological advancements can enhance resilience to climatic challenges. Today, continued innovation in areas such as renewable energy, climate-resilient infrastructure, and sustainable agriculture is essential for adapting to the impacts of climate change.

- **Community Involvement:** Many of the successful adaptation strategies from history were community-driven, involving collective action and cooperation. The construction and maintenance of irrigation systems, the mobility of nomadic tribes, and the communal efforts in terrace farming all relied on strong community involvement.

Ancient Civilizations and Water Management

One of the earliest and most significant examples of economic adaptation to climate variations can be found in ancient civilizations such as Mesopotamia, Egypt, and the Indus Valley. These societies developed advanced water management systems to cope with variable climate conditions, demonstrating their resilience and ingenuity in the face of environmental challenges.

Mesopotamia: The Cradle of Civilization

Mesopotamia, often referred to as the "Cradle of Civilization", was located in the fertile crescent between the Tigris and Euphrates rivers. This region experienced unpredictable and often severe flooding, which posed a significant threat to agricultural production and settlements. To address these challenges, the Mesopotamians constructed intricate irrigation canals, levees, dikes, and reservoirs. These water management systems were designed to control the flow of the rivers, store water during periods of excess, and distribute it during dry spells. The construction of these irrigation systems required substantial communal effort and advanced engineering knowledge.

Canals diverted water from the rivers to fields, enabling the cultivation of crops in areas that would otherwise be arid. Reservoirs stored excess water from seasonal floods, providing a crucial resource during droughts. Levees and dikes protected fields and settlements from destructive floods. The success of Mesopotamian agriculture, bolstered by these irrigation systems, led to surplus production, which in turn supported the growth of cities and the development of complex societies. The ability to manage water resources effectively allowed Mesopotamia to become one of the earliest and most influential civilizations in human history.

Egypt: The Gift of the Nile

Ancient Egypt is another prime example of economic adaptation through advanced water management. The Nile River, which flows through one of the driest regions on Earth, provided a lifeline for the Egyptian civilization.

The Nile's annual flooding deposited nutrient-rich silt on its banks, creating fertile land for agriculture. However, the timing and extent of the floods varied, posing challenges for consistent agricultural production. The Egyptians developed extensive irrigation systems to harness the benefits of the Nile's floods while mitigating their unpredictability. They constructed canals, basins, and dikes to control and distribute floodwaters across their fields. These irrigation networks allowed them to extend cultivation beyond the immediate floodplain, increasing the amount of arable land.

One notable innovation was the basin irrigation system, where water was diverted into large basins and then released gradually to irrigate crops. This method maximized water use efficiency and minimized the risk of crop failure during periods of low rainfall. The construction of these irrigation systems required organized labour and centralized control, reflecting the sophisticated social and administrative structures of ancient Egypt. The ability to manage the Nile's waters enabled Egypt to maintain stable agricultural production, supporting a large population and the development of one of the most enduring civilizations in history. The surplus food production facilitated trade, economic growth, and the accumulation of wealth, contributing to Egypt's status as a major power in the ancient world.

Indus Valley: Urban Planning and Water Management

The Indus Valley Civilization, located in present-day Pakistan and north-west India, is another example of early economic adaptation to climate variations through advanced water management. The cities of Harappa and Mohenjo-Daro, among others, displayed remarkable urban planning and hydraulic engineering. The Indus Valley people constructed sophisticated drainage systems, wells, and reservoirs to manage water resources and address the challenges of both flooding and drought. Their cities were built with a grid layout, incorporating advanced sanitation and drainage systems that directed wastewater away from residential areas.

This not only improved public health but also reflected a high level of urban planning and social organization. Irrigation was essential for agriculture in the Indus Valley, where rainfall was seasonal and unpredictable. The civilization developed a network of canals to divert river water to their fields, ensuring a reliable supply of water for crops. Reservoirs and storage tanks were built to collect and store rainwater, providing a crucial resource during dry periods.

The success of these water management systems contributed to the prosperity and stability of the Indus Valley Civilization. Agricultural surplus supported urban centres, trade, and a complex economy. The ability to adapt to climate variations through effective water management underscores the ingenuity and resilience of the Indus Valley people.

Conclusion

The early examples of economic adaptation to climate variations in ancient Mesopotamia, Egypt, and the Indus Valley highlight the critical role of advanced water management systems in ensuring stable agricultural production and supporting complex societies. These civilizations developed innovative solutions to manage water resources, demonstrating their resilience in the face of environmental challenges. In Mesopotamia, intricate irrigation canals and reservoirs controlled unpredictable river flooding, enabling surplus production and urban growth. In Egypt, extensive irrigation systems harnessed the Nile's annual floods, maximizing agricultural output and supporting a powerful civilization.

In the Indus Valley, sophisticated drainage and irrigation systems facilitated urban planning and agricultural stability. These historical examples provide valuable lessons for modern societies facing the impacts of climate change. The principles of innovation, communal effort, and sustainable resource management that underpinned ancient water management systems remain relevant today. By studying and learning from these early adaptations, contemporary efforts to address climate challenges can be informed and improved, ensuring resilience and sustainability for future generations.

Nomadic Pastoralism: Adapting to Arid Conditions

In regions where sedentary agriculture was less viable due to harsh, arid conditions, such as the Arabian Peninsula and parts of Central Asia, communities developed a unique form of adaptation known as nomadic pastoralism. This lifestyle relied on the mobility of livestock herds to access grazing lands and water sources across vast territories, enabling these societies to survive and thrive in challenging environments by moving in response to seasonal variations and resource availability.

The Arabian Peninsula: Survival in the Desert

The Bedouin tribes of the Arabian Peninsula epitomize the adaptive strategies of nomadic pastoralism. Living in one of the most arid and inhospitable regions on Earth, these tribes developed a deep understanding of their environment and a way of life that revolved around the mobility of their livestock, primarily camels, sheep, and goats.

- **Mobility and Resource Management:** The Bedouins' survival depended on their ability to move their herds to different grazing areas based on the availability of water and forage.

This mobility was meticulously planned and executed, taking advantage of their extensive knowledge of the desert landscape, seasonal patterns, and the locations of oases and temporary water sources. By moving their herds, the Bedouins could avoid overgrazing any single area, allowing vegetation to recover and ensuring a sustainable supply of food for their animals.

- **Social Organization and Economy:** The Bedouin society was structured around tribal affiliations, with each tribe possessing detailed knowledge of specific territories. This knowledge was passed down through generations, enabling them to navigate the desert and manage their resources effectively. The economy of the Bedouins was primarily based on the trade of livestock and livestock products, such as milk, meat, and wool, with settled communities. This trade provided them with essential goods, such as grain and manufactured items that were not available in their nomadic lifestyle.
- **Cultural Practices and Resilience:** The cultural practices of the Bedouins, including their social norms and traditions, were intrinsically linked to their nomadic way of life. Hospitality, for instance, was a crucial aspect of Bedouin culture, ensuring mutual support and cooperation among tribes in the harsh desert environment. This strong sense of community and mutual aid enhanced their resilience, allowing them to endure the hardships of desert living.

Central Asia: The Steppes and Nomadic Empires

In Central Asia, the vast steppes provided a different but equally challenging environment for nomadic pastoralists. Tribes such as the Mongols developed a nomadic lifestyle that enabled them to thrive in the region's extreme climates, characterized by hot summers and brutally cold winters.

- **Seasonal Migration:** Central Asian nomads practised seasonal migration, known as transhumance, moving their herds of horses, sheep, and goats between summer and winter pastures.

In the summer, they moved to higher altitudes where the temperatures were cooler and grazing was abundant. In the winter, they descended to lower altitudes to escape the harsh cold. This movement ensured that their livestock had access to fresh pastures and reduced the risk of overgrazing.

- **Political and Military Power:** The mobility and organizational skills of Central Asian nomads also had significant political and military implications.

  The ability to move quickly across vast distances gave these nomadic tribes a strategic advantage in warfare. The Mongol Empire, under the leadership of Genghis Khan, utilized these skills to create one of the largest empires in history. The economic base of this empire was supported by the productivity of their pastoralist lifestyle and the wealth generated through trade and conquest.
- **Economic Integration:** The nomads of Central Asia were integral to the economic networks of the Silk Road, connecting East and West through trade. They facilitated the exchange of goods, ideas, and technologies across continents. Their herds provided valuable commodities such as wool, leather, and meat, which were traded for silk, spices, and other luxury items. This economic integration not only supported their livelihoods but also contributed to the prosperity of the regions they connected.

Lessons from Nomadic Pastoralism

The historical examples of nomadic pastoralism in the Arabian Peninsula and Central Asia offer valuable lessons for contemporary adaptation strategies. These societies' ability to thrive in harsh climates was based on several key principles that remain relevant today.

- **Mobility and Flexibility:** The nomadic pastoralists' success was rooted in their mobility and flexibility. By moving in response to environmental conditions, they could optimize resource use and avoid the pitfalls of overexploitation. Modern adaptation strategies can benefit from similar flexibility, incorporating adaptive management practices that respond to changing environmental conditions.

- **Knowledge and Tradition:** The deep environmental knowledge and cultural traditions of nomadic societies were crucial to their resilience. This highlights the importance of incorporating local knowledge and practices into contemporary adaptation efforts. Engaging with and learning from indigenous and traditional communities can provide valuable insights and enhance the effectiveness of adaptation strategies.
- **Sustainable Resource Use:** Nomadic pastoralists practised sustainable resource use, balancing the needs of their herds with the capacity of the environment. This sustainable approach is essential for modern adaptation, emphasizing the need to manage resources in a way that ensures their long-term availability. By adopting practices that promote environmental stewardship and resource conservation, contemporary societies can mitigate the adverse impacts of climate change and build resilient, sustainable economies.

Conclusion

Nomadic pastoralism in regions such as the Arabian Peninsula and Central Asia exemplifies how human societies have historically adapted to harsh climates through mobility, flexibility, and sustainable resource management. These adaptive strategies allowed nomadic communities to survive and thrive in environments where sedentary agriculture was not viable. The lessons learned from their practices provide valuable insights for contemporary adaptation efforts, highlighting the importance of mobility, local knowledge, and sustainability in developing effective responses to climate challenges. By drawing on the principles of nomadic pastoralism, modern societies can enhance their resilience and adaptability in the face of a changing climate. This historical perspective underscores the enduring relevance of traditional adaptation strategies and their potential to inform and inspire current and future climate resilience initiatives.

Lessons from Historical Adaptation Strategies

Analysing historical adaptation strategies provides several lessons that can inform current efforts to address climate change. These lessons highlight the importance of innovation, community involvement, and sustainable practices in developing effective adaptation measures.

- **Innovation and Technological Development:** Historical examples demonstrate that innovation and the development of new technologies have been crucial for successful adaptation.

  Ancient water management systems, terrace farming, and other technological advancements enabled societies to overcome environmental challenges and enhance their resilience. For instance, the sophisticated irrigation systems of ancient Mesopotamia, which included canals and levees, allowed for the control of water flow and minimized the risks posed by flooding and drought. Similarly, the Inca civilization's terrace farming in the Andes Mountains maximized arable land use and improved agricultural productivity despite challenging terrain and variable climate conditions.

  Today, the development and deployment of modern technologies are essential for adapting to contemporary climate risks. Renewable energy systems, such as solar and wind power, reduce dependency on fossil fuels and mitigate the impacts of climate change. Advanced agricultural practices, including precision farming and genetically modified crops, enhance food security by increasing crop yields and resilience to extreme weather. Innovations in building climate-resilient infrastructure, such as flood barriers and sustainable urban planning, also play a critical role in protecting communities from climate-related disasters.

- **Community Involvement and Local Knowledge:** Successful historical adaptation strategies often involved the active participation of local communities and the use of indigenous knowledge. This community-based approach ensured that adaptation measures were culturally appropriate, socially acceptable, and effectively implemented. For example, the traditional agricultural practices of the Zuni people in the south-west America included the construction of waffle gardens—small, grid-like plots surrounded by earthen walls that retained moisture and reduced soil erosion.

  These practices, rooted in local knowledge, allowed the Zuni to cultivate crops in arid conditions and sustain their communities. In contemporary adaptation efforts, community involvement and the integration of local knowledge remain crucial. Engaging local communities in the planning and implementation of adaptation measures ensures that these initiatives are tailored to the specific needs and conditions of the area. It also fosters a sense of ownership and responsibility among community members, enhancing the sustainability and effectiveness of adaptation projects. Collaborating with indigenous peoples and respecting their traditional ecological knowledge can provide valuable insights and innovative solutions to climate challenges.
- **Sustainable Resource Management:** Historical adaptation strategies often emphasized the sustainable management of natural resources. Ancient societies understood the importance of balancing resource use with environmental capacity to ensure long-term survival. The Mayans, for instance, practised agro-forestry—a method of integrating trees and shrubs into agricultural landscapes—to maintain soil fertility, conserve water, and protect against soil

erosion. This approach allowed them to sustainably manage their agricultural systems and support large populations.

Modern adaptation strategies must similarly prioritize sustainable resource management. Practices such as water conservation, reforestation, and sustainable agriculture are vital for maintaining ecosystem health and resilience. Implementing integrated water resource management (IWRM) approaches, which consider the interconnections between land, water, and ecosystems, can help balance competing demands and ensure the sustainable use of water resources. Promoting sustainable agriculture practices, such as organic farming and agro-ecology, can enhance soil health, increase biodiversity, and reduce greenhouse gas emissions.

- **Flexibility and Adaptive Management:** Historical examples show that flexibility and the ability to adapt to changing conditions are key components of successful adaptation strategies. Nomadic pastoralists, for example, exemplified adaptive management by moving their herds in response to seasonal variations and resource availability. This flexibility allowed them to optimize resource use and avoid overexploitation, ensuring the sustainability of their livelihoods.

  In contemporary adaptation efforts, adopting flexible and adaptive management practices are essential. Climate change is characterized by uncertainty and variability, necessitating approaches that can respond dynamically to changing conditions. Adaptive management involves monitoring environmental and social conditions, learning from experience, and adjusting strategies as needed. This iterative process enables continuous improvement and enhances the resilience of adaptation measures.
- **Integration of Multiple Approaches:** Historical adaptation strategies often integrated multiple approaches to address complex environmental challenges. For instance, the

ancient Egyptians combined advanced water management systems with agricultural innovations, such as crop rotation and fertilization, to ensure food security and resilience to climate variability. This holistic approach addressed multiple aspects of their society and environment, creating a robust and adaptable system.

Modern adaptation strategies can similarly benefit from an integrated approach. Combining technological innovations, community-based initiatives, sustainable resource management, and adaptive management practices can create comprehensive and effective adaptation measures. For example, urban areas can integrate green infrastructure, such as parks and wetlands, with grey infrastructure, such as flood barriers and drainage systems, to manage storm water and reduce flood risks. Agricultural systems can combine traditional knowledge with modern techniques to enhance resilience and productivity.

- **Policy and Governance:** Effective policy and governance frameworks have historically played a crucial role in supporting adaptation strategies. The management of communal resources, enforcement of regulations, and provision of incentives were essential for the successful implementation of adaptation measures. In ancient China, for example, the central government played a key role in coordinating large-scale water management projects, such as the construction of the Grand Canal, which facilitated irrigation and trade.

  Today, strong policy and governance frameworks are equally important for addressing climate change. Governments at all levels must create enabling environments for adaptation through policies that promote sustainable practices, provide funding and support for adaptation projects, and ensure the participation of all stakeholders. International cooperation and agreements, such as the Paris Agreement, also play a vital role in fostering global efforts to combat climate change.

Conclusion

The lessons from historical adaptation strategies highlight the importance of innovation, community involvement, sustainable resource management, flexibility, integrated approaches, and effective governance in addressing climate change. By learning from these historical examples, modern societies can develop robust and resilient adaptation measures that enhance their ability to cope with contemporary climate challenges. These lessons underscore the need for a holistic and inclusive approach to climate adaptation, leveraging both traditional knowledge and modern technologies to build a sustainable and resilient future.

Community Involvement and Collective Action

Many historical adaptation strategies were community-driven, involving collective action and cooperation among members of society. These strategies highlight the importance of community involvement in successfully adapting to environmental challenges, offering valuable insights for contemporary climate adaptation efforts.

Historical Examples of Community-Driven Adaptation

Ancient societies often relied on collective action to implement and maintain adaptation strategies. For instance, the success of ancient irrigation systems in regions like Mesopotamia, Egypt, and the Indus Valley was heavily dependent on coordinated community efforts. These complex networks of canals, dams, and reservoirs required regular maintenance and management, tasks that were often organized and executed by local communities. The collective responsibility for these systems ensured their functionality and sustainability, allowing agricultural productivity to be maintained even in the face of variable climatic conditions.

In the Andean region, the Inca civilization developed sophisticated agricultural terraces that were constructed and maintained by local communities. These terraces prevented soil erosion, conserved water, and allowed for the cultivation of crops at different altitudes. The communal labour system, known as "mit'a", facilitated the construction and upkeep of these terraces, demonstrating the critical role of collective action in sustaining agricultural practices in challenging environments.

Modern Adaptation Strategies and Community Involvement

Contemporary climate adaptation strategies can greatly benefit from community involvement and collective action. Engaging communities in planning and decision-making processes ensures that adaptation measures are context-specific, culturally appropriate, and widely accepted. This participatory approach not only enhances the effectiveness of adaptation initiatives but also fosters a sense of ownership and responsibility among community members.

Local Knowledge and Expertise

Incorporating local knowledge and expertise is a key aspect of community-driven adaptation. Indigenous and local communities possess a deep understanding of their environment, built over generations of interaction with their natural surroundings. This knowledge can provide valuable insights into sustainable resource management and adaptation practices. For example, in the Pacific Islands, traditional knowledge about weather patterns, marine resources, and coastal processes, has been crucial in developing adaptation strategies to cope with rising sea levels and increasing storm intensity.

Community-led initiatives, such as the construction of traditional seawalls and the restoration of mangroves, have proven effective in enhancing coastal resilience. In Africa, the use of indigenous knowledge in managing water resources has been instrumental in adapting to changing climatic conditions. In the Sahel region, traditional rainwater harvesting techniques, such as the construction of "zaï" pits and "half-moons", have been revitalized to improve soil fertility and agricultural productivity. These techniques, developed and refined over centuries, are being integrated into modern adaptation strategies to enhance food security and resilience to drought.

Participatory Approaches and Social Cohesion

Participatory approaches to adaptation involve actively engaging community members in the design, implementation, and monitoring of adaptation measures. This inclusive process ensures that the diverse needs and perspectives of the community are considered, leading to more effective and equitable outcomes. In Bangladesh, community-based adaptation strategies have been employed to address the impacts of flooding and sea-level rise. Local communities have been involved in constructing raised housing, floating gardens, and cyclone shelters, which have improved resilience and provided economic opportunities.

The participatory approach has strengthened social cohesion and empowered communities to take proactive measures in managing climate risks. In urban areas, participatory approaches have been used to develop climate-resilient infrastructure and enhance community resilience. In the city of Chennai, India, residents have been engaged in the restoration of urban wetlands and the implementation of rainwater harvesting systems. These initiatives have not only reduced flood risks but also improved water security and urban biodiversity.

Challenges and Opportunities

While community involvement and collective action offer numerous benefits, they also present challenges. Ensuring meaningful participation can be difficult, particularly in communities with diverse and sometimes conflicting interests. Effective communication, trust-building, and capacity-building are essential to overcoming these challenges and fostering successful collaboration. Moreover, community-driven adaptation requires adequate support from governments, non-governmental organizations, and other stakeholders. Providing financial resources, technical assistance, and policy support can enhance the capacity of communities to implement and sustain adaptation measures.

Conclusion

Historical examples of community-driven adaptation underscore the importance of collective action and local knowledge in managing environmental challenges. These lessons are highly relevant for contemporary climate adaptation efforts, highlighting the need for participatory approaches that engage communities in planning and decision-making processes. By leveraging the expertise and involvement of local communities, modern adaptation strategies can be more context-specific, culturally appropriate, and widely accepted, leading to more effective and sustainable outcomes. The integration of community involvement and collective action into climate adaptation not only enhances resilience but also fosters social cohesion and empowerment. As climate change continues to pose significant challenges, the lessons from historical adaptation strategies and the active participation of communities will be crucial in developing robust and inclusive responses.

Sustainable Practices

Historical adaptation strategies often incorporated sustainable practices that balanced resource use with environmental conservation. These practices ensured the long-term viability of resources and the health of ecosystems, allowing societies to thrive over extended periods. Understanding these sustainable practices provides valuable insights for modern adaptation efforts, which must prioritize sustainability to ensure long-term resilience and avoid exacerbating climate-related vulnerabilities.

Terrace Farming: A Model of Sustainable Agriculture

Terrace farming is a prime example of a historical adaptation strategy that promoted sustainability. This agricultural technique, used in various parts of the world, involved creating stepped terraces on mountain slopes to cultivate crops. The Incas in the Andes, the Ifugao in the Philippines, and the Chinese in Yunnan are notable practitioners of this method.

- **Preventing Soil Erosion:** One of the primary benefits of terrace farming is its ability to prevent soil erosion. The terraces act as barriers, slowing down water runoff and allowing it to infiltrate the soil, reducing the loss of topsoil. This is particularly important in mountainous regions where heavy rains can wash away fertile soil, rendering the land barren and unproductive.
- **Promoting Biodiversity:** Terrace farming also promotes biodiversity. The creation of multiple micro-environments along the terraces supports a variety of crops and plant species. This diversity not only enhances food security by providing a range of food sources but also contributes to ecological stability. Different plant species can support various insects, birds, and other wildlife, fostering a balanced ecosystem.
- **Maintaining Soil Fertility:** Terraces help maintain soil fertility by preventing nutrient runoff and allowing organic matter to decompose and enrich the soil. The practice often includes crop rotation and the use of organic fertilizers, further enhancing soil health. These methods ensure that the soil remains productive over long periods, supporting sustainable agricultural practices.

Agro-forestry: Integrating Trees and Agriculture

Agro-forestry, the integration of trees and shrubs into agricultural landscapes, is another historical practice that embodies sustainability. This method has been used in regions such as the Sahel in Africa, the Amazon Basin, and South-east Asia.

- **Enhancing Soil Quality:** Trees and shrubs in agro-forestry systems contribute to soil quality by adding organic matter through leaf litter and root decomposition. They also help fix nitrogen in the soil, improving its fertility and reducing the need for chemical fertilizers.
- **Conserving Water:** Trees play a crucial role in conserving water. Their root systems enhance soil structure, increasing

its capacity to retain water. This is particularly beneficial in arid and semi-arid regions where water is scarce. By maintaining soil moisture levels, agro-forestry systems help sustain crops during dry periods.

- **Providing Multiple Products:** Agro-forestry systems offer a diverse range of products, including fruits, nuts, timber, and medicinal plants, in addition to traditional crops. This diversity supports economic resilience by providing farmers with multiple income sources. It also reduces pressure on forest resources, as communities can obtain wood and other products from their agro-forestry plots rather than from natural forests.

Rotational Grazing: Sustainable Livestock Management

Rotational grazing is a sustainable livestock management practice that involves moving animals between different pasture areas to prevent overgrazing and allow vegetation to recover.

This method has been used by pastoralist communities around the world, including the Maasai in East Africa and the nomadic tribes of Central Asia.

- **Preventing Overgrazing:** Rotational grazing prevents overgrazing by ensuring that no single area is continuously exposed to grazing pressure. This allows grasses and other plants to regenerate, maintaining healthy pastures and reducing soil erosion.
- **Improving Soil Health:** This practice also improves soil health. As animals graze, they naturally fertilize the land with their manure, which decomposes and enriches the soil with nutrients. The rest periods between grazing allow the soil to recover and maintain its structure and fertility.
- **Enhancing Biodiversity:** Rotational grazing supports biodiversity by promoting a variety of plant species. Different plants have different growth rates and nutrient needs, and rotating grazing areas allow a diverse mix of vegetation to thrive. This diversity provides a habitat for

various insects, birds, and other wildlife, contributing to a balanced ecosystem.

Modern Implications and Practices

The sustainable practices observed in historical adaptation strategies provide essential lessons for contemporary efforts to address climate change. Modern adaptation strategies must incorporate sustainability to ensure long-term resilience and avoid exacerbating climate-related vulnerabilities.

- **Water Conservation:** Like ancient terrace farming, modern agricultural practices can prioritize water conservation. Techniques such as drip irrigation, rainwater harvesting, and the use of drought-resistant crop varieties can help manage water resources sustainably. These methods reduce water waste, enhance soil moisture, and ensure that crops receive adequate water even during dry periods.
- **Soil Health:** Maintaining soil health is crucial for sustainable agriculture. Practices such as cover cropping, reduced tillage, and organic farming can enhance soil fertility and structure. These methods increase organic matter in the soil, improve its ability to retain water, and reduce erosion, ensuring long-term productivity.
- **Biodiversity:** Promoting biodiversity is essential for ecological stability and resilience. Modern agricultural systems can integrate practices such as agro-forestry, polycultures, and permaculture to support diverse plant and animal species. This diversity enhances ecosystem services, such as pollination, pest control, and nutrient cycling, which are vital for sustainable food production.
- **Community-based Approaches:** Community involvement in sustainable practices is critical for their success. Engaging local communities in planning and implementing adaptation measures ensures that these initiatives are context-specific and culturally appropriate. It also fosters

a sense of ownership and responsibility, enhancing the sustainability of these practices.

- **Policy Support:** Effective policies and governance are essential for promoting sustainable practices. Governments can provide incentives for sustainable farming, invest in research and development of sustainable technologies, and enforce regulations that protect natural resources. International cooperation and agreements can also play a vital role in supporting sustainable adaptation strategies.

Conclusion

Historical adaptation strategies often incorporated sustainable practices that balanced resource use with environmental conservation. Terrace farming, agro-forestry, and rotational grazing are examples of how ancient societies managed their resources sustainably to ensure long-term resilience. These practices offer valuable lessons for modern adaptation efforts, emphasizing the importance of water conservation, soil health, biodiversity, community involvement, and policy support. By prioritizing sustainability, contemporary adaptation strategies can enhance resilience and mitigate the adverse impacts of climate change, ensuring a sustainable future for generations to come.

Comparative Analysis of Past and Present Adaptation Methods

Comparing historical and contemporary adaptation methods reveals both continuities and changes in how societies respond to climate challenges. While the context and scale of adaptation efforts have evolved, some underlying principles remain consistent.

Scale and Complexity

One of the key differences between past and present adaptation methods is the scale and complexity of the challenges and responses. Historical adaptations were often localized, addressing specific climatic variations within a limited geographical area. For example, ancient irrigation systems in Mesopotamia or terrace farming in the

Andes were tailored to address the unique environmental conditions of those regions.

These methods were effective for managing localized challenges such as flooding or soil erosion but were not designed to address broader or interconnected climatic issues. In contrast, modern climate change poses global challenges that require coordinated responses at multiple levels, from local to international. The impacts of climate change are widespread, affecting ecosystems, economies, and societies across the globe. This necessitates a more comprehensive and integrated approach to adaptation.

Contemporary adaptation efforts involve advanced technologies, sophisticated modelling, and comprehensive policy frameworks that consider the interconnected nature of global climate systems. For instance, the Paris Agreement represents a global effort to address climate change, with countries committing to reduce greenhouse gas emissions and enhance resilience through nationally determined contributions (NDCs). This level of international coordination and commitment was not present in historical adaptation efforts, highlighting the increased scale and complexity of modern responses.

Technological Advancements

Technological advancements have significantly transformed adaptation methods over time. Historical societies relied on relatively simple but innovative technologies to manage their environmental challenges. For example, the construction of terraces for agriculture in the Andean region or the use of *qanats* (underground water channels) in ancient Persia demonstrated sophisticated engineering within the context of their time. Today's adaptation strategies leverage cutting-edge technologies that offer unprecedented capabilities. Remote sensing, geographic information systems (GIS), and climate modelling enable precise monitoring and prediction of climate impacts.

Renewable energy technologies, such as solar and wind power, reduce dependency on fossil fuels and mitigate climate change.

Advanced agricultural practices, including precision farming and genetically modified crops, enhance food security and resilience to extreme weather. Furthermore, modern infrastructure projects incorporate climate resilience into their design. For instance, the Netherlands' Delta Works, a series of sophisticated flood protection systems, integrate advanced engineering and real-time monitoring to protect against sea-level rise and storm surges. Such technological integration was not possible in historical contexts but exemplifies how contemporary adaptations benefit from technological progress.

Policy and Governance

Effective policy and governance frameworks are crucial for both historical and contemporary adaptation efforts, though the scope and mechanisms have evolved. Historical adaptations often involved local governance and communal efforts. For example, the management of water resources in ancient Egypt relied on a centralized authority to oversee irrigation projects and ensure equitable water distribution. In modern times, adaptation policies and governance structures are more complex, involving multiple levels of government, international organizations, and a wide range of stakeholders. Policies such as the European Union's Green Deal or national adaptation plans (NAPs) outline comprehensive strategies for enhancing climate resilience.

These frameworks often include regulations, incentives, and funding mechanisms to support adaptation initiatives across various sectors. International cooperation is also a key feature of contemporary adaptation efforts. Organizations like the United Nations Framework Convention on Climate Change (UNFCCC) facilitate global dialogue and cooperation on climate issues. The Global Environment Facility (GEF) and the Green Climate Fund (GCF) provide financial support to developing countries for implementing adaptation projects. Such international mechanisms reflect the collaborative nature of modern adaptation strategies, which was less prominent in historical contexts.

Sustainability and Community Involvement

Sustainability and community involvement are principles that underpin both historical and contemporary adaptation methods, though their application has evolved. Historical adaptations often emphasized sustainable practices that balanced resource use with environmental conservation. For example, the Inca's terrace farming maintained soil fertility and prevented erosion, ensuring long-term agricultural productivity. Modern adaptation strategies continue to prioritize sustainability, but with a broader understanding of ecological interdependencies and the need for systemic change. Sustainable practices now encompass renewable energy, sustainable agriculture, and green infrastructure. Additionally, contemporary approaches recognize the importance of social sustainability, ensuring that adaptation measures are inclusive and equitable. Community involvement remains crucial for successful adaptation.

Historical adaptations were often community-driven, relying on collective action and local knowledge. This is still relevant today, as engaging local communities in planning and decision-making processes ensure that adaptation measures are context-specific and culturally appropriate. Programmes such as community-based adaptation (CBA) emphasize participatory approaches, leveraging local knowledge and fostering community ownership of adaptation projects.

Integration of Multidisciplinary Approaches

Another key difference between past and present adaptation methods is the integration of multidisciplinary approaches. Historical adaptations often focused on specific challenges, such as managing water resources or improving agricultural practices. While these efforts were effective within their scope, they did not always address the broader impacts of environmental changes on society and the economy. Modern adaptation strategies benefit from a multidisciplinary approach that integrates climate science, economics, social sciences, and engineering. This holistic perspective allows for comprehensive solutions that address the multifaceted nature of climate impacts. For instance, urban planning now incorporates climate resilience into infrastructure design, transportation systems, and public health initiatives, creating more resilient cities.

Conclusion

Comparing historical and contemporary adaptation methods reveals both continuities and changes in how societies respond to climate challenges. While the scale and complexity of modern adaptation efforts have increased, the underlying principles of innovation, sustainability, community involvement, and effective governance remain consistent. Technological advancements, multidisciplinary approaches, and international cooperation characterize contemporary adaptation strategies, reflecting the global nature of current climate challenges. By learning from historical successes and leveraging modern capabilities, societies can develop robust and effective adaptation measures to address the ongoing and future impacts of climate change.

Technological Advancements

The technological capabilities available today far surpass those of historical societies, enabling more effective and diverse adaptation strategies.

Modern technologies, such as satellite monitoring, climate modelling, and renewable energy systems, provide detailed data and innovative solutions that were unimaginable in the past. However, the principles of leveraging technology and innovation to enhance resilience remain consistent.

Advanced Monitoring and Prediction Tools

One of the most significant technological advancements in modern adaptation strategies is the development of advanced monitoring and prediction tools. Satellite technology and remote sensing allow for real-time monitoring of climate variables, such as temperature, precipitation, and sea-level changes. These tools provide detailed and accurate data, enabling scientists and policymakers to track climate trends, forecast extreme weather events, and assess vulnerabilities. Climate modelling is another critical advancement. Using sophisticated computer simulations, scientists can predict future climate scenarios based on different greenhouse gas emission pathways. These models help identify potential impacts on various sectors, such as agriculture, water resources, and infrastructure, allowing for proactive adaptation planning. Historical societies lacked such predictive capabilities, relying instead on empirical observations and traditional knowledge to guide their adaptation efforts.

Renewable Energy Systems

The development and deployment of renewable energy systems represent a major technological breakthrough in addressing climate change. Renewable energy sources, such as solar, wind, and hydroelectric power, offer sustainable alternatives to fossil fuels, reducing greenhouse gas emissions and mitigating climate change. The integration of renewable energy into national grids and the expansion of decentralized energy systems, such as solar panels and wind turbines, enhance energy security and resilience. Historically, societies relied on renewable energy in more rudimentary forms, such as windmills for grinding grain or water wheels for irrigation. While effective within their context, these technologies could not match the scale and efficiency of modern renewable energy systems. Today's advancements allow for large-scale energy production and distribution, supporting sustainable development and reducing dependency on fossil fuels.

Climate-Resilient Infrastructure

Modern technological advancements have also revolutionized the design and construction of climate-resilient infrastructure.

Engineering innovations enable the development of buildings, roads, bridges, and other structures that can withstand extreme weather events and climate-related hazards. For example, flood-resistant building designs, seawalls, and storm surge barriers protect coastal cities from rising sea levels and storm surges. Historical societies also implemented innovative infrastructure solutions, such as the stepped wells of India or the *qanat* systems in Persia, to manage water resources and mitigate the impacts of droughts and floods. However, the scale and complexity of modern infrastructure projects coupled with advanced materials and engineering techniques, provide a higher level of resilience and adaptability to contemporary climate challenges.

Sustainable Agriculture

Technological advancements in agriculture have significantly enhanced the resilience of food systems to climate change. Precision farming techniques, such as drip irrigation, soil moisture sensors, and GPS-guided machinery, optimize water and nutrient use, improving crop yields and reducing resource waste. Genetic engineering and biotechnology have led to the development of climate-resilient crop varieties that can tolerate extreme temperatures, drought, and pests. Historically, societies developed various agricultural innovations, such as terrace farming in the Andes or crop rotation practices in medieval Europe, to adapt to local environmental conditions. These practices laid the foundation for sustainable agriculture, emphasizing soil conservation and biodiversity. Modern advancements build on these principles, incorporating cutting-edge technology to address the growing challenges of climate change and global food security.

Information and Communication Technology

Information and communication technology (ICT) plays a crucial role in modern adaptation strategies. The widespread availability of mobile phones, the internet, and social media facilitates the dissemination of climate information, early warning systems, and disaster preparedness plans. ICT tools enable real-time communication and coordination among governments, communities, and international organizations, enhancing the efficiency and effectiveness of adaptation efforts. Historically, the transmission of climate information relied on traditional communication methods, such as oral traditions, community meetings, and early warning signals.

While effective within their contexts, these methods lacked the speed and reach of modern ICT tools. Today, digital platforms and mobile applications provide instant access to critical information, empowering communities to respond quickly to climate-related threats.

Conclusion

The technological advancements available today provide unprecedented capabilities for climate adaptation, offering innovative solutions and detailed data that were unimaginable in historical contexts. Advanced monitoring and prediction tools, renewable energy systems, climate-resilient infrastructure, sustainable agriculture, and information and communication technology have transformed adaptation strategies, enhancing resilience and sustainability. However, the underlying principles of leveraging technology and innovation to address environmental challenges remain consistent across time. By integrating modern advancements with traditional knowledge and sustainable practices, societies can develop robust and effective adaptation measures to address the ongoing and future impacts of climate change.

Policy and Governance

The role of policy and governance in shaping adaptation efforts represents another significant difference between historical and contemporary approaches. Historical adaptations were often driven by local practices and community initiatives, with limited involvement from centralized authorities. In contrast, contemporary adaptation requires robust policy frameworks, regulatory support, and international cooperation. Governments, intergovernmental organizations, and NGOs play a crucial role in coordinating and implementing adaptation strategies, ensuring that they are integrated into broader development plans and aligned with global climate goals.

Historical Adaptation and Local Governance

In historical contexts, adaptation strategies were largely community-driven, reflecting the decentralized nature of governance at the time. Local communities, relying on traditional knowledge and practices, developed and implemented adaptation measures to cope with their specific environmental challenges. These efforts were typically small-scale and tailored to local conditions. For example, ancient water management systems in Mesopotamia and the Indus Valley were built and maintained by local communities. These systems, including irrigation canals and reservoirs, were essential for agricultural productivity and were managed collectively.

Similarly, in the Andean region, the construction and upkeep of agricultural terraces were organized through communal labour systems, such as the Incan "mit'a", which distributed the workload among community members. While effective within their limited scope, these historical adaptation efforts were constrained by the lack of centralized coordination and comprehensive policy frameworks. Decisions were made at the local level, often without a broader understanding of regional or global climatic trends, which limited the scalability and integration of these strategies.

Contemporary Adaptation and Centralized Governance

In contrast, contemporary adaptation efforts are characterized by the involvement of centralized authorities and the development of comprehensive policy frameworks. Governments at various levels, intergovernmental organizations, and NGOs play a crucial role in shaping, coordinating, and implementing adaptation strategies. These efforts are often integrated into broader development plans and aligned with global climate goals.

- **National Policies and Frameworks:** Modern adaptation strategies are often guided by national policies and frameworks that provide a coordinated approach to addressing climate change. For instance, many countries have developed National Adaptation Plans (NAPs) that outline their priorities and actions for enhancing climate resilience. These plans are informed by scientific research and stakeholder consultations, ensuring that they are comprehensive and context-specific.

- **Regulatory Support:** Effective adaptation requires robust regulatory support to enforce standards and incentivize sustainable practices. Governments implement regulations to ensure that infrastructure projects, agricultural practices, and industrial activities are climate-resilient and environmentally sustainable. For example, building codes may be updated to require flood-resistant construction in vulnerable areas, and agricultural policies may promote water-efficient irrigation techniques and climate-resilient crop varieties.
- **International Cooperation:** The global nature of climate change necessitates international cooperation. Organizations such as the United Nations Framework Convention on Climate Change (UNFCCC) facilitate collaboration among countries to address climate impacts. The Paris Agreement, a landmark international accord, commits countries to limit global warming and enhance adaptive capacities.
- **Role of NGOs and Civil Society:** Non-governmental organizations (NGOs) and civil society play a vital role in supporting adaptation efforts. NGOs often act as intermediaries, bridging the gap between local communities and governments. They provide technical expertise, raise awareness, and mobilize resources for adaptation projects. For instance, organizations like the Red Cross and Red Crescent Societies implement community-based adaptation programmes, focusing on building resilience at the grassroots level.

- **Integrated Approaches:** Contemporary adaptation strategies emphasize the integration of climate resilience into broader development agendas. This involves mainstreaming adaptation into sectors such as agriculture, water resources, infrastructure, and public health. Integrated approaches ensure that adaptation measures are not implemented in isolation but are part of comprehensive development plans that address multiple objectives.

Challenges and Opportunities

While the involvement of centralized authorities and the development of comprehensive policies offer significant advantages, they also present challenges. Ensuring effective coordination among multiple stakeholders, including government agencies, NGOs, and local communities, can be complex. Policy coherence and alignment are crucial to avoid fragmented efforts and ensure that adaptation strategies are synergistic and mutually reinforcing. Moreover, the implementation of adaptation policies often faces barriers such as limited financial resources, political will, and institutional capacity.

Developing countries, in particular, may struggle to mobilize the necessary funding and technical expertise to implement their adaptation plans. International support and cooperation are essential to address these gaps and enable effective adaptation at the local and national levels. However, the structured and coordinated approach of contemporary adaptation efforts also presents opportunities. By leveraging scientific research and technological advancements, policymakers can design evidence-based strategies that are both effective and scalable. The integration of climate resilience into national development plans ensures that adaptation measures contribute to sustainable development goals, promoting economic growth, social equity, and environmental sustainability.

Conclusion

The role of policy and governance in shaping adaptation efforts marks a significant evolution from historical to contemporary approaches. While historical adaptations were often driven by local practices and community initiatives, modern adaptation requires robust policy frameworks, regulatory support, and international cooperation. Governments, intergovernmental organizations, and NGOs play a crucial role in coordinating and implementing adaptation strategies, ensuring that they are integrated into broader development plans and aligned with global climate goals. By learning from the strengths and limitations of historical adaptation efforts, contemporary strategies can leverage centralized coordination, scientific research, and international collaboration to enhance resilience and sustainability. Effective governance and policy support are essential for addressing the complex and interconnected challenges posed by climate change, ensuring that adaptation efforts are comprehensive, inclusive, and impactful.

Sustainability and Equity

While historical adaptation strategies often emphasized sustainability, modern efforts must also address issues of equity and social justice. Climate change disproportionately affects vulnerable populations, exacerbating existing inequalities. Ensuring that adaptation measures are inclusive and equitable is a critical challenge that requires targeted interventions, capacity building, and support for marginalized communities.

Historical Emphasis on Sustainability

Historical adaptation strategies frequently prioritized sustainable practices that balanced resource use with environmental conservation. For instance, the agricultural terraces of the Inca civilization in the Andes promoted soil fertility, prevented erosion, and maintained ecological balance. Similarly, the water management systems of ancient Egypt and Mesopotamia were designed to optimize resource use while ensuring long-term agricultural productivity. These practices highlight an understanding of sustainability that was crucial for the survival and prosperity of ancient societies.

Modern Challenges of Equity and Social Justice

In contemporary times, the challenge of climate adaptation extends beyond sustainability to include equity and social justice. Climate change impacts are not evenly distributed; they often hit the poorest and most vulnerable populations hardest.

These groups, which include low-income communities, indigenous peoples, women, children, and the elderly, typically have the least capacity to adapt and recover from climate-related shocks.

- **Targeted Interventions:** Ensuring equity in climate adaptation requires targeted interventions that address the specific needs of vulnerable populations. For instance, in coastal regions prone to sea-level rise, building protective infrastructure such as seawalls and elevated housing can safeguard the homes and livelihoods of low-income communities. In agriculture, introducing drought-resistant crop varieties and providing access to irrigation systems can help smallholder farmers adapt to changing rainfall patterns.
- **Capacity Building:** Building the capacity of marginalized communities to adapt to climate change is essential for achieving equity. This includes providing education, training, and resources that enable these communities to develop and implement their own adaptation strategies. For example, community-based adaptation programmes often involve training local people in sustainable agricultural practices, water management, and disaster preparedness. These programmes empower communities to take proactive measures and enhance their resilience to climate impacts.
- **Support for Marginalized Communities:** Financial and technical support for marginalized communities is critical. Governments, international organizations, and NGOs can provide funding, technology, and expertise to help these communities implement effective adaptation measures. For

instance, the Green Climate Fund (GCF) supports projects that promote climate resilience in developing countries, with a focus on vulnerable populations. Such support helps bridge the gap between the needs of marginalized communities and the resources required to address them.

Knowledge and Information Sharing

The dissemination and exchange of knowledge have become more sophisticated and widespread in the modern era. Historical societies relied on local knowledge and traditions to develop adaptation strategies. Today, the global exchange of scientific research, best practices, and innovative solutions facilitates the rapid dissemination of information and fosters collaboration across borders.

Historical Reliance on Local Knowledge

In the past, societies adapted to climate variability by relying on accumulated local knowledge and traditional practices. For example, indigenous communities often developed detailed understanding of their local ecosystems, which guided their agricultural, water management, and settlement practices. This knowledge was passed down through generations, ensuring that adaptation measures were well-suited to local conditions and sustainably managed.

Modern Global Exchange of Knowledge

In contemporary times, the exchange of knowledge has expanded dramatically due to advances in communication technologies and increased international cooperation. The internet, mobile technology, and satellite communications enable the rapid dissemination of climate data, scientific research, and adaptation strategies worldwide. This global exchange of knowledge enhances the ability of societies to learn from each other and implement effective adaptation measures.

- **Scientific Research and Best Practices:** The global scientific community plays a crucial role in advancing our understanding of climate change and developing adaptation strategies. Research institutions, universities, and international organizations conduct studies that provide insights into the impacts of climate change and the effectiveness of various adaptation measures. These findings are shared through academic publications, conferences, and online platforms, making them accessible to policymakers, practitioners, and the public.

- **Innovative Solutions:** The exchange of innovative solutions across borders fosters collaboration and innovation in climate adaptation. For example, the use of satellite technology for monitoring climate variables, the development of climate-resilient infrastructure, and the implementation of renewable energy systems are innovations that have been shared globally. International cooperation through initiatives such as the Intergovernmental Panel on Climate Change (IPCC) and the United Nations Framework Convention on Climate Change (UNFCCC) facilitates the dissemination of these solutions and encourages countries to adopt best practices.
- **Capacity Building and Training:** Knowledge sharing also involves capacity building and training programmes that equip individuals and communities with the skills needed to adapt to climate change. International organizations and NGOs often conduct workshops, training sessions, and educational programmes to transfer knowledge and build local capacity. These efforts ensure that communities are not only aware of the available adaptation measures but also capable of implementing them effectively.

Conclusion

The principles of sustainability and equity are integral to modern adaptation efforts, reflecting a broader understanding of the interconnectedness between environmental health and social justice. While historical adaptation strategies often emphasized sustainable practices, contemporary efforts must address the additional challenge of ensuring that adaptation measures are inclusive and equitable. This requires targeted interventions, capacity building, and support for marginalized communities. The sophisticated and widespread dissemination of knowledge in the modern era further enhances our ability to respond to climate challenges. By leveraging global scientific research, sharing best practices, and fostering international collaboration, societies can develop and implement effective adaptation strategies.

The combination of sustainability, equity, and knowledge sharing provides a comprehensive framework for building resilience in the face of climate change, ensuring that adaptation efforts are both effective and just. Historical perspectives on economic adaptation to climate change reveal a wealth of knowledge and experience that can inform contemporary efforts to address climate challenges. Early examples of adaptation, such as ancient water management systems, nomadic pastoralism, and terrace farming, demonstrate the importance of innovation, community involvement, and sustainable practices. Lessons from these historical strategies highlight the need for technological advancement, collective action, and equitable solutions in modern adaptation efforts. By comparing past and present methods, we can identify enduring principles and adapt them to the current context, ensuring that our responses to climate change are effective, inclusive, and sustainable.

REFERENCES

Auerbach, A.J., and Gorodnichenko, Y. (2022). "Fiscal Multipliers in Recessions and Expansions: Does It Matter Whether Government Spending is Increasing or Decreasing?" *Journal of Economic Perspectives, 36*(1): 41-64.

Ball, L.M., Gagnon, E., Honohan, P. and Krogstrup, S. (2022). "What Else Can Central Banks Do?" *IMF Economic Review, 70*(1): 167-206.

Bianchi, F. and Melosi, L. (2022). "Inflation Expectations and the Supply Chain". *American Economic Review, 112*(4): 1284-1320.

Blanchard, O. (2023). "The Policy Toolkit for Inflation: Interest Rates, Fiscal Policy, and Supply Side Measures". *Journal of Policy Modeling, 45*(2): 203-217.

Bordo, M.D. and Levin, A.T. (2022). "Central Bank Digital Currency and the Future of Monetary Policy". *National Bureau of Economic Research Working Paper*, No. 29923.

Christiano, L.J., Eichenbaum, M.S. and Trabandt, M. (2022). "Understanding the Great Inflation: Lessons for Current Policy". *Quarterly Journal of Economics, 137*(3): 1263-1300.

Clarida, R.H. (2022). "Monetary Policy, Price Stability, and Inflation: Theoretical Perspectives and Practical Lessons". *Journal of Economic Literature, 60*(1): 34-62.

Coibion, O., Gorodnichenko, Y. and Weber, M. (2022). "How Inflation Expectations Affect Households and Firms: The Role of Central Bank Communication". *Brookings Papers on Economic Activity, 2022*(1): 115-174.

Cogley, T. and Sbordone, A.M. (2022). "Inflation and the Evolution of Price Dispersion". *Journal of Monetary Economics, 122*: 156-170.

Daly, M.C. and Hobijn, B. (2022). "Wage Growth and Inflation: The Role of Productivity and Labour Market Slack". *Economic Review, Federal Reserve Bank of Kansas City, 2022*(Q2): 47-74.

Draghi, M. (2022). "Policy Coordination in a High-Inflation Environment". *Journal of International Economics*, Vol. 132, Editorial Board Article No. 103594.

Drechsler, I., Savov, A. and Schnabl, P. (2022). "The Financial Stability Implications of Inflation". *Journal of Finance, 77*(4): 1873-1911.

Eggertsson, G.B. and Woodford, M. (2022). "The Zero Bound on Interest Rates and Optimal Monetary Policy". *Brookings Papers on Economic Activity, 2022*(2): 139-211.

Gertler, M. and Karadi, P. (2022). "Monetary Policy Surprises, Credit Costs, and Economic Activity". *American Economic Journal: Macroeconomics, 14*(2): 78-117.

Goodhart, C.A.E. and Pradhan, M. (2022). "The Great Demographic Reversal and Inflation". *Journal of Economic Dynamics and Control*, Vol. 135, Editorial Board Article No. 104297.

Gourinchas, P.O. (2022). "Global Economic Challenges: Inflation, Trade, and Policy". *IMF Economic Review, 70*(3): 249-278.

Guvenen, F., Kaplan, G. and Song, J. (2022). "How Does Inflation Affect Different Income Groups?" *Journal of Political Economy, 130*(5): 1437-1473.

Hamilton, J.D. (2022). "Why You Should Never Use the Hodrick-Prescott Filter". *Review of Economic Studies, 89*(1): 104-120.

Jordà, Ò. and Taylor, A.M. (2022). "The Pandemic, Inflation, and Debt Sustainability". *Economic Policy, 37*(110): 275-314.

3

The Economics of Climate Risk Assessment and Management

Overview of Climate Risk Assessment Methodologies

Climate risk assessment involves evaluating the potential impacts of climate change on various sectors and regions, identifying vulnerabilities, and prioritizing actions to mitigate risks. These assessments are crucial for understanding the economic implications of climate change and for developing effective risk management strategies. This process can be broken down into several key methodologies: hazard identification and analysis, vulnerability assessment, and impact assessment.

Hazard Identification and Analysis

The first step in climate risk assessment is identifying and analyzing climate hazards. This involves recognizing the types of extreme weather events and long-term climatic changes that could potentially impact a region. Examples of these hazards include floods, hurricanes, droughts, sea-level rise, and temperature increases.

- **Historical Data:** Analyzing historical data helps establish patterns and frequencies of past climate events. This data provides a baseline for understanding the extent and severity of previous hazards, which can inform predictions about future events. Historical records of temperature, precipitation, storm occurrences, and sea levels are invaluable for this analysis.
- **Climate Models:** Climate models are essential tools for predicting future climate conditions. These models simulate the Earth's climate system and project changes based on different greenhouse gas emission scenarios. They provide detailed forecasts of temperature, precipitation, storm intensity, and sea-level rise, among other variables. The outputs from these models help identify areas that are likely to be affected by various climate hazards.
- **Geographical Information Systems (GIS):** GIS technology is used to map and analyze the spatial distribution of climate hazards. By overlaying climate model projections onto maps of physical and human geography, GIS helps identify areas at risk from specific hazards. For example, GIS can highlight flood-prone regions, coastal areas vulnerable to sea-level rise, and zones susceptible to extreme heat.

Hazard identification and analysis set the stage for understanding where and how climate impacts may occur, providing a foundation for further risk assessment steps.

Vulnerability Assessment

Once climate hazards are identified, the next step is to assess the vulnerability of different sectors, communities, and ecosystems to these hazards. Vulnerability assessments determine how susceptible various elements are to climate impacts, considering factors such as socio-economic status, infrastructure robustness, and environmental conditions.

- **Socio-economic Status:** Vulnerability is often higher among poorer communities due to limited resources and adaptive capacity. Socio-economic factors, such as income levels, education, and access to healthcare, influence a community's ability to prepare for, respond to, and recover from climate impacts. For instance, low-income households may lack the financial means to reinforce their homes against floods or access alternative sources of water during droughts.
- **Infrastructure Robustness:** The condition and resilience of infrastructure play a significant role in vulnerability. Critical infrastructure such as roads, bridges, water supply systems, and power grids must withstand climate impacts. Areas with outdated or poorly maintained infrastructure are more vulnerable to damage and disruption from extreme weather events.
- **Environmental Conditions:** Natural ecosystems can either buffer or amplify climate impacts. Healthy ecosystems, such as wetlands and forests, can absorb excess water, reduce flood risks, and mitigate temperature extremes. Conversely, degraded environments are less capable of providing these services, increasing vulnerability to climate hazards.
- **Vulnerability and Capacity Assessment (VCA) Framework:** Tools like the VCA framework help identify and map vulnerable populations and assets. This framework involves a systematic approach to assess vulnerabilities and capacities at the community level, incorporating local knowledge and stakeholder input. By engaging with communities, the VCA framework ensures that the assessment is context-specific and considers the unique characteristics and needs of the population.

Conducting a thorough vulnerability assessment is crucial for identifying which sectors, communities, and ecosystems require the most urgent attention and intervention.

It highlights where resources should be directed to enhance resilience and reduce susceptibility to climate impacts.

Impact Assessment

Following the identification of hazards and assessment of vulnerabilities, the next step is to evaluate the potential impacts of climate change. Impact assessment involves quantifying the consequences of climate hazards on various aspects of society and the environment.

- **Sectoral Impact Analysis:** This involves examining how climate hazards affect different sectors such as agriculture, water resources, health, and infrastructure. For example, in agriculture, extreme weather events like droughts and floods can reduce crop yields and livestock productivity, leading to food insecurity and economic losses. In the health sector, heatwaves can increase the incidence of heat-related illnesses and exacerbate existing health conditions.
- **Economic Impact:** Assessing the economic impacts of climate change involves estimating the costs associated with damage and disruption. This includes direct costs such as repair and reconstruction of damaged infrastructure, as well as indirect costs like loss of income, increased healthcare expenses, and decreased productivity. Economic impact assessments help quantify the financial burden of climate change and justify investments in adaptation measures.
- **Social Impact:** Climate change can have profound social impacts, including displacement, loss of livelihoods, and exacerbation of inequalities. Assessing social impacts involves understanding how different groups are affected, with a focus on vulnerable populations such as women, children, and the elderly and indigenous communities. Social impact assessments help ensure that adaptation measures are inclusive and address the needs of those most at risk.

- **Environmental Impact:** Environmental impact assessments evaluate how climate hazards affect natural ecosystems. This includes changes in biodiversity, ecosystem services, and habitat integrity. Understanding environmental impacts is crucial for developing conservation and restoration strategies that enhance ecosystem resilience.

Impact assessments provide a comprehensive understanding of the potential consequences of climate change, informing the development of targeted and effective adaptation measures.

Prioritizing Actions

The final step in climate risk assessment is prioritizing actions to mitigate risks. This involves identifying the most critical vulnerabilities and impacts, and determining which adaptation measures will provide the greatest benefits.

- **Risk Matrix:** A risk matrix is a tool used to prioritize actions by plotting the likelihood and severity of different climate impacts. This visual representation helps decision-makers focus on high-risk areas and allocate resources accordingly.
- **Cost-benefit Analysis (CBA):** CBA is used to compare the costs of implementing adaptation measures against the anticipated benefits in terms of reduced damages and enhanced resilience. It helps identify the most economically efficient strategies by quantifying both costs and benefits in monetary terms.
- **Stakeholder Engagement:** Engaging stakeholders in the prioritization process ensures that adaptation measures are context-specific and widely accepted. This includes involving local communities, government agencies, NGOs, and the private sector in decision-making and implementation.
- **Implementation and Monitoring:** Once priorities are set, the next step is to implement adaptation measures and establish monitoring systems to track their effectiveness.

Continuous monitoring and evaluation allow for adjustments based on new information and changing conditions, ensuring that adaptation efforts remain relevant and effective over time.

Conclusion

Climate risk assessment is a multifaceted process that involves identifying hazards, assessing vulnerabilities, evaluating impacts, and prioritizing actions to mitigate risks. By employing robust methodologies such as hazard identification and analysis, vulnerability assessment, and impact assessment, policymakers and stakeholders can develop comprehensive strategies to enhance resilience and reduce the economic and social costs of climate change. Ensuring that these efforts are inclusive and equitable is crucial for addressing the disproportionate impacts on vulnerable populations. Through informed decision-making and coordinated action, it is possible to build a more resilient and sustainable future in the face of climate change.

Impact Assessment: Understanding the Consequences of Climate Hazards

Impact assessment is a critical component of climate risk assessment, focusing on evaluating the potential consequences of climate hazards on various aspects of society. This step involves a comprehensive analysis of how different sectors, including economic activities, public health, infrastructure, and natural resources, may be affected by climate change. To achieve this, impact assessments utilize climate models, economic models, and scenario analysis to estimate potential damages and disruptions.

Economic Impact

- **Sectoral Analysis:** One of the primary areas of focus in impact assessment is the economic impact of climate hazards. This involves examining how climate change affects various economic sectors, such as agriculture, industry, and services. For instance, agriculture is highly sensitive to changes in temperature, precipitation, and extreme weather events. Droughts can lead to crop failures, reducing yields and income for farmers, while floods can destroy crops and infrastructure, leading to significant economic losses. Similarly, the industrial sector may face disruptions in production due to damage to infrastructure and supply chains, while the services sector, including tourism and hospitality, may experience reduced demand due to adverse climate conditions.

- **Cost Estimation:** Economic impact assessments also involve estimating the costs associated with climate hazards. These costs include direct costs, such as the repair and reconstruction of damaged infrastructure, and indirect costs, such as loss of productivity, increased healthcare expenses, and reduced economic output. By quantifying these costs, policymakers can understand the financial burden of climate change and justify investments in adaptation measures.

Public Health Impact

- **Health Risks:** Climate change poses significant risks to public health and impact assessments aim to quantify these risks. Extreme weather events, such as heat waves, floods, and storms, can lead to immediate health impacts, including injuries, heat-related illnesses, and fatalities. Moreover, long-term changes in climate can exacerbate existing health conditions and increase the prevalence of diseases.

  For instance, rising temperatures can expand the range of vector-borne diseases like malaria and dengue fever, while changes in precipitation patterns can affect the spread of water-borne diseases.
- **Vulnerability Analysis:** Public health impact assessments consider the vulnerability of different populations to

climate hazards. Vulnerable groups, such as the elderly, children, and low-income communities, are often at higher risk of adverse health outcomes due to limited access to healthcare and adaptive resources. By identifying these vulnerable populations, impact assessments can inform targeted interventions to protect public health and enhance community resilience.

Infrastructure Impact

- **Damage and Disruption:** Infrastructure is crucial for the functioning of society, and its vulnerability to climate hazards is a major concern. Impact assessments evaluate the potential damage to critical infrastructure, including transportation networks, energy systems, water supply, and sanitation facilities. Extreme weather events can cause physical damage to infrastructure, leading to disruptions in essential services. For example, hurricanes and floods can damage roads, bridges, and power lines, disrupting transportation and electricity supply. Such disruptions can have cascading effects on economic activities and public health, highlighting the importance of resilient infrastructure.
- **Adaptation and Resilience:** Assessing the impact of climate hazards on infrastructure also involves evaluating the effectiveness of adaptation measures. This includes analyzing the resilience of existing infrastructure and identifying opportunities for improvement. For instance, impact assessments may examine the potential benefits of upgrading infrastructure to withstand extreme weather events, such as reinforcing buildings, elevating roads, and enhancing drainage systems. These measures can reduce the vulnerability of infrastructure and minimize disruptions, contributing to overall societal resilience.

Natural Resources Impact

- **Ecosystem Services:** Natural resources and ecosystems provide essential services that support human well-being, including water supply, food production, and climate regulation. Impact assessments evaluate how climate hazards affect these ecosystem services. For example, changes in temperature and precipitation can alter water availability, affecting both human consumption and agricultural irrigation.
  Similarly, extreme weather events can damage forests, wetlands, and coastal ecosystems, reducing their capacity to provide services such as flood protection and carbon sequestration.
- **Biodiversity:** Climate change can also have profound impacts on biodiversity, with implications for ecosystem health and resilience. Impact assessments analyze how climate hazards affect species distribution, population dynamics, and habitat integrity. For instance, rising temperatures and changing precipitation patterns can shift the distribution of plant and animal species, leading to changes in ecosystem composition and function. Understanding these impacts is crucial for developing conservation strategies that protect biodiversity and maintain ecosystem services.

Scenario Analysis

- **Exploring Futures:** Scenario analysis is a key tool in impact assessments, allowing researchers to explore different future climate conditions and their potential impacts. By considering a range of scenarios, including varying levels of greenhouse gas emissions and socio-economic development pathways, impact assessments can provide insights into the potential consequences of different policy choices. This approach helps policymakers understand the range of possible outcomes and make informed decisions about adaptation and mitigation strategies.

- **Uncertainty and Risk:** Scenario analysis also addresses the inherent uncertainty in climate projections and impact assessments. By considering multiple scenarios, researchers can identify the range of possible impacts and assess the likelihood of different outcomes. This helps in developing robust adaptation strategies that are effective under various future conditions, reducing the risk of maladaptation and ensuring long-term resilience.

Conclusion

Impact assessment is a vital step in climate risk assessment, providing a comprehensive understanding of the potential consequences of climate hazards on economic activities, public health, infrastructure, and natural resources. By utilizing climate models, economic models, and scenario analysis, impact assessments can estimate potential damages and disruptions, informing the development of targeted and effective adaptation measures. Ensuring that these measures address the needs of vulnerable populations and enhance overall resilience is crucial for mitigating the impacts of climate change and building a sustainable future.

Risk Characterization: Quantifying the Overall Risk of Climate Change

Risk characterization is a crucial step in the climate risk assessment process, as it combines hazard, vulnerability, and impact assessments to quantify the overall risk posed by climate change. This step translates complex scientific data into actionable information by expressing risk in terms of the probability and magnitude of potential impacts. By doing so, it provides a basis for prioritizing risk management actions and allocating resources effectively.

Integrating Hazard, Vulnerability, and Impact Assessments

- **Hazard Assessments:** The first component of risk characterization involves understanding the likelihood and nature of climate hazards. This includes analyzing the frequency, intensity, and duration of extreme weather events such as floods, hurricanes, droughts, and heat waves. Climate models and historical data play a crucial role in predicting these hazards, offering insights into future climate scenarios and their potential impacts.

- **Vulnerability Assessments:** Vulnerability assessments identify which sectors, communities, and ecosystems are most susceptible to climate hazards. This involves evaluating socio-economic factors, infrastructure resilience, and environmental conditions that influence the ability to cope with and recover from climate impacts. For instance, low-income communities often have fewer resources to invest in adaptive measures, making them more vulnerable to extreme weather events.
- **Impact Assessments:** The impact assessment component estimates the potential consequences of climate hazards on various aspects of society, including economic activities, public health, infrastructure, and natural resources. This involves quantifying both direct and indirect impacts, such as property damage, loss of productivity, increased healthcare costs, and ecosystem degradation.

By integrating these three assessments, risk characterization provides a comprehensive picture of climate risk, highlighting where and how climate change could cause significant disruptions.

Quantifying Risk: Probability and Magnitude

- **Probability of Impacts:** One of the key elements in risk characterization is determining the probability of different climate impacts. This involves using statistical methods and climate models to estimate the likelihood of various hazards occurring under different climate scenarios.
  For example, models can predict the probability of a 100-year flood event occurring more frequently due to increased precipitation patterns. Understanding these probabilities

helps in planning for and mitigating the most likely and severe climate impacts.

- **Magnitude of Impacts:** In addition to probability, the magnitude of potential impacts is crucial for characterizing risk. This involves assessing the scale and severity of consequences associated with different climate hazards. For instance, the magnitude of impact from a major hurricane can include extensive infrastructure damage, significant economic losses, and long-term displacement of communities. Quantifying these impacts in terms of economic costs, human health effects, and environmental degradation helps prioritize areas that need urgent intervention.
- **Risk Matrices:** Risk matrices are commonly used tools in risk characterization that combine probability and magnitude to categorize risks into different levels of severity. A risk matrix typically uses a grid format, with one axis representing the probability of an event and the other axis representing the magnitude of its impact. This visual representation helps stakeholders easily identify high-risk areas that require immediate attention and lower-risk areas that may need less urgent but still important interventions.

Prioritizing Risk Management Actions

- **Identifying High Risk Areas:** By quantifying risks, stakeholders can identify which areas and sectors are at the highest risk and prioritize them for intervention. For example, coastal cities facing high probabilities of sea-level rise and storm surges can be prioritized for investments in protective infrastructure, such as seawalls and flood barriers. Similarly, agricultural regions vulnerable to drought can be targeted for water management improvements and the introduction of drought-resistant crop varieties.
- **Resource Allocation:** Effective risk characterization informs resource allocation by identifying where investments will

have the greatest impact. Limited resources can be directed towards high-risk areas to maximize the effectiveness of adaptation measures. For instance, funding for climate resilience projects can be prioritized for communities with high vulnerability and significant expected impacts, ensuring that resources are used efficiently and effectively.

- **Developing Adaptation Strategies:** Quantifying risk provides a solid foundation for developing targeted adaptation strategies. This involves designing specific measures to reduce vulnerability and enhance resilience based on the identified risks. For example, urban planning can integrate climate risk data to avoid developing in high-risk flood zones, while healthcare systems can prepare for the increased likelihood of heat-related illnesses by expanding cooling centres and emergency response capabilities.
- **Monitoring and Review:** Risk characterization is not a one-time process; it requires ongoing monitoring and review. Climate conditions and socio-economic factors can change, altering the risk landscape. Continuous monitoring allows for the updating of risk assessments and adaptation strategies to reflect new information and emerging risks. This dynamic approach ensures that risk management actions remain relevant and effective over time.

Communication and Stakeholder Engagement

- **Effective Communication:** Communicating the results of risk characterization to stakeholders is essential for informed decision-making. This involves translating complex scientific data into clear and understandable information that can be used by policymakers, community leaders, businesses, and the public. Visual tools such as maps, graphs, and risk matrices are particularly useful for conveying the spatial and temporal dimensions of climate risk.

- **Stakeholder Engagement:** Engaging stakeholders in the risk characterization process enhances its relevance and acceptance. Involving local communities, businesses, government agencies, and NGOs ensures that diverse perspectives are considered and that the resulting adaptation strategies are inclusive and equitable. Stakeholder engagement also fosters collaboration and collective action, which are critical for effective climate risk management.

Conclusion

Risk characterization is a vital step in the climate risk assessment process, integrating hazard, vulnerability, and impact assessments to quantify the overall risk posed by climate change. By expressing risk in terms of probability and magnitude, this step provides a basis for prioritizing risk management actions and allocating resources effectively.

Through effective communication and stakeholder engagement, risk characterization ensures that adaptation strategies are informed, targeted, and inclusive, ultimately enhancing resilience and reducing the impacts of climate change on society.

Economic Models for Risk Management

Economic models play a crucial role in analyzing the costs and benefits of different risk management strategies and in integrating climate risks into economic planning. These models provide a framework for evaluating the economic implications of climate change and help policymakers make informed decisions about adaptation and mitigation efforts. Understanding the various types of economic models used in risk management is essential for developing effective strategies to address climate risks.

Types of Economic Models

Cost-Benefit Analysis (CBA)

- **Cost-Benefit Analysis (CBA):** CBA is a fundamental economic model used to compare the costs of implementing adaptation measures against the benefits derived from reduced climate risks. CBA helps determine the most economically efficient strategies by quantifying both costs and benefits in monetary terms. This approach allows policymakers to prioritize actions that provide the greatest net benefits to society.
- **Steps in Conducting CBA:** The first step in conducting CBA is identifying all potential adaptation options. These options can range from infrastructure investments, such as building seawalls, to policy measures, such as implementing water conservation regulations. Once the options are identified, the next step is to estimate the costs associated with each option. These costs include initial capital expenditures, operation and maintenance costs, and any indirect costs such as social or environmental impacts. The benefits of each adaptation option are then quantified in monetary terms. Benefits can include avoided damages from climate hazards, such as reduced flood damage, lower healthcare costs due to fewer heat-related illnesses, and increased agricultural productivity from improved water management. The present value of these benefits is calculated using a discount rate to account for the time value of money.

  Finally, the net present value (NPV) of each option is determined by subtracting the present value of costs from the present value of benefits. Adaptation measures with positive NPVs are considered economically viable, while those with negative NPVs may be less attractive. Sensitivity analysis is often conducted to test the robustness of the CBA results by varying key assumptions and parameters, such as discount rates and cost estimates.

Integrated Assessment Models (IAMs)

Integrated Assessment Models (IAMs) combine climate science, economics, and policy analysis to evaluate the interactions between climate change and economic systems. IAMs are used to assess the long-term economic impacts of climate change and the effectiveness of different mitigation and adaptation policies.

- **Structure of IAMs:** IAMs integrate data from various disciplines to provide a comprehensive view of climate change impacts and policy responses. These models typically include modules for climate dynamics, economic growth, energy systems, and land use. By simulating how changes in one area affect others, IAMs help policymakers understand the trade-offs and synergies between different policy options.
- **Applications of IAMs:** IAMs are used to evaluate the costs and benefits of global climate policies, such as carbon pricing, emissions reduction targets, and renewable energy investments. They also help in assessing the economic impacts of climate change on different regions and sectors, providing valuable insights for national and international policy development.

Computable General Equilibrium (CGE) Models

Computable General Equilibrium (CGE) models simulate how economies respond to changes in policies, prices, and external shocks, such as climate impacts. CGE models are used to assess the broader economic effects of climate risk management strategies, including impacts on GDP, employment, and sectoral outputs.

- **Structure of CGE Models:** CGE models are based on the principles of supply and demand in a market economy. They include multiple sectors, households, and government entities, capturing the interactions between different parts of the economy. CGE models use mathematical equations to represent production, consumption, trade,

and investment decisions, allowing for a detailed analysis of how changes in one area affect the entire economy.
- **Applications of CGE Models:** CGE models are used to analyze the economic impacts of climate policies, such as carbon taxes, subsidies for renewable energy, and adaptation investments. They help policymakers understand the distributional effects of these policies, identifying which sectors and regions are most affected and how resources can be allocated to mitigate adverse impacts.

Dynamic Stochastic General Equilibrium (DSGE) Models

Dynamic Stochastic General Equilibrium (DSGE) models incorporate random shocks and dynamic adjustments to analyze the macroeconomic implications of climate risks and policy responses over time. DSGE models are particularly useful for understanding how economies evolve under different climate scenarios and policy interventions.

- **Structure of DSGE Model:** DSGE models extend the CGE framework by incorporating stochastic elements, such as random climate shocks, and dynamic processes, such as investment and technological change. These models use forward-looking equations to capture the expectations and behaviour of economic agents, providing a more realistic representation of economic dynamics.
- **Applications of DSGE Models:** DSGE models are used to simulate the long-term effects of climate policies, such as investments in green technology and infrastructure. They help policymakers evaluate the resilience of the economy to climate shocks and the effectiveness of different policy measures in promoting sustainable growth and reducing vulnerability to climate risks.

Conclusion

Economic models are essential tools for analyzing the costs and benefits of different risk management strategies and integrating climate risks into economic planning. Models such as Cost-Benefit Analysis (CBA), Integrated Assessment Models (IAMs), Computable General Equilibrium (CGE) models, and Dynamic Stochastic General Equilibrium (DSGE) models provide valuable insights into the economic implications of climate change and the effectiveness of adaptation and mitigation policies. By leveraging these models, policymakers can develop informed, economically efficient strategies to manage climate risks and promote sustainable development.

Integrated Assessment Models (IAMs)

Integrated Assessment Models (IAMs) are sophisticated tools that combine climate science, economics, and policy analysis to evaluate the interactions between climate change and economic systems. These models are instrumental in assessing the long-term economic impacts of climate change and the effectiveness of various mitigation and adaptation policies. By integrating diverse data sets and disciplines, IAMs provide a comprehensive framework for understanding the multifaceted nature of climate change and for informing policy decisions.

Structure and Components of IAMs

IAMs integrate multiple components to create a holistic view of the climate-economy nexus. The main components typically include:

- **Climate Dynamics:** This module incorporates scientific data on climate processes, such as greenhouse gas emissions, atmospheric concentrations, radioactive forcing, and temperature changes. It often utilizes outputs from climate models to simulate future climate scenarios under different emissions pathways.
- **Economic Growth:** This component models the economy, including sectors such as energy, industry, agriculture, and services. It examines how economic activities generate emissions and how these emissions affect economic output and growth. Economic models within IAMs often include

production functions, consumption patterns, trade, and investment dynamics.

- **Energy Systems:** The energy module assesses the supply and demand of various energy sources, including fossil fuels, renewable, and nuclear power. It evaluates the costs and potentials of different energy technologies and their impacts on emissions and economic growth. This module is critical for exploring the transition to low-carbon energy systems.
- **Land Use and Agriculture:** This component examines land use changes, agricultural productivity, deforestation, and reforestation. It assesses the impacts of climate change on food security and land-based carbon sequestration. By including land use, IAMs can evaluate the trade-offs between agriculture, forestry, and bio-energy production.
- **Policy Analysis:** IAMs incorporate policy instruments such as carbon pricing, emissions trading systems, subsidies for renewable energy, and regulations on emissions. This module evaluates the economic and environmental impacts of different policy scenarios, helping to identify effective strategies for mitigation and adaptation.

Applications of IAMs

IAMs are used for various applications, providing valuable insights into the complex interactions between climate change and socio-economic systems.

- **Evaluating Climate Policies:** One of the primary applications of IAMs is to assess the effectiveness of climate policies. For instance, IAMs can evaluate the impacts of implementing a global carbon tax or an emissions trading system. By simulating different policy scenarios, IAMs help policymakers understand the potential economic and environmental outcomes, such as emission reductions, changes in GDP, and shifts in energy consumption patterns.

- **Long-term Economic Impacts:** IAMs provide long-term projections of the economic impacts of climate change. They estimate the costs associated with climate damages, such as reduced agricultural yields, increased healthcare expenses, and damage to infrastructure. By comparing these costs with the benefits of mitigation and adaptation measures, IAMs help identify economically viable strategies for addressing climate change.
- **Assessing Mitigation Pathways:** IAMs are used to explore various mitigation pathways and their feasibility. For example, IAMs can simulate scenarios where renewable energy sources gradually replace fossil fuels, or where carbon captures and storage (CCS) technologies are widely deployed. These simulations help assess the technological, economic, and policy requirements for achieving specific climate targets, such as limiting global warming to 1.5°C or 2°C above pre-industrial level.
- **Integrating Adaptation Measures:** IAMs also integrate adaptation measures into their analyses. This includes evaluating the costs and benefits of adaptation strategies, such as building flood defences, improving water management, and enhancing agricultural resilience. By assessing both mitigation and adaptation, IAMs provide a comprehensive view of how societies can respond to climate change.
- **Global and Regional Analysis:** IAMs can operate at global, regional, and national scales, providing insights tailored to different geographical contexts. This flexibility allows for detailed analyses of specific regions or countries, considering their unique economic structures, vulnerabilities, and policy environments. For example, IAMs can evaluate the impacts of climate policies on developing countries, identifying opportunities for sustainable development and resilience-building.

Examples of IAMs

Several well-known IAMs have been developed and are widely used in climate research and policy analysis:

- **The DICE Model (Dynamic Integrated Climate Economy):** Developed by William Nordhaus, the DICE model integrates economic and climate dynamics to assess the optimal path for carbon pricing and emissions reductions. It is used to estimate the social cost of carbon and to evaluate the economic benefits of climate policies.
- **The FUND Model (Framework for Uncertainty, Negotiation, and Distribution):** The FUND model focuses on the distributional impacts of climate change and policies. It assesses how different regions and income groups are affected by climate change and explores equity considerations in climate policy.
- **The MESSAGE Model (Model for Energy Supply Strategy Alternatives and their General Environmental Impact):** Developed by the International Institute for Applied Systems Analysis (IIASA), MESSAGE focuses on energy systems and their interactions with the economy and environment. It is used to explore energy transition scenarios and their implications for emissions and economic development.

Benefits and Limitations

- **Benefits:** IAMs provide a comprehensive framework for understanding the complex interactions between climate change and socio-economic systems. They offer valuable insights for policymakers by integrating scientific, economic, and policy data. IAMs help identify cost-effective strategies for mitigation and adaptation, supporting informed decision-making.
- **Limitations:** Despite their strengths, IAMs have limitations. They rely on numerous assumptions and

simplifications, which can introduce uncertainties. The accuracy of IAM projections depends on the quality of input data and the robustness of the underlying models. Additionally, IAMs may not fully capture the socio-political dynamics and behavioural responses that influence climate policy outcomes.

Conclusion

Integrated Assessment Models (IAMs) are powerful tools that combine climate science, economics, and policy analysis to evaluate the interactions between climate change and economic systems.

By providing long-term projections and assessing the effectiveness of various mitigation and adaptation policies, IAMs play a crucial role in informing climate policy decisions. Despite their limitations, IAMs offer valuable insights that help policymakers develop strategies to address the economic and environmental challenges posed by climate change, ultimately contributing to sustainable development and resilience-building efforts worldwide.

Computable General Equilibrium (CGE) Models

Computable General Equilibrium (CGE) models are essential tools in economic analysis, particularly in assessing the broad economic effects of changes in policies, prices, and external shocks such as climate impacts. These models simulate how economies respond to various changes, providing insights into the impacts on GDP, employment, sectoral outputs, and overall economic welfare.

Structure and Functioning of CGE Models

- **Economic Representation:** CGE models represent the economy as a system of interrelated markets for goods, services, and factors of production (such as labour and capital). They are based on the principles of supply and demand, where markets are assumed to clear (i.e. supply equals demand) through price adjustments.

- **Equilibrium Conditions:** The core concept of CGE models is equilibrium, where all markets in the economy are in balance. The models use mathematical equations to describe the behaviour of economic agents (households, firms, government) and the interactions between them. The equilibrium conditions ensure that the total supply of each good or service equals total demand.
- **Data and Calibration:** CGE models are calibrated using a comprehensive set of economic data, typically derived from national accounts, input-output tables, and social accounting matrices. This data provides a snapshot of the economy at a specific point in time, which the model uses as a baseline for simulations.

Applications of CGE Models

- **Policy Analysis:** CGE models are widely used to analyze the economic impacts of policy changes. For example, they can simulate the effects of introducing a carbon tax on different sectors of the economy. By capturing the interactions between sectors, CGE models can show how a carbon tax might shift production and consumption patterns, affect prices, and influence overall economic performance.
- **Climate Risk Management:** In the context of climate risk management, CGE models help assess the broader economic effects of adaptation and mitigation strategies. For instance, they can evaluate the economic benefits of investing in flood defences or the impacts of adopting renewable energy technologies. These models provide valuable insights into the trade-offs and synergies between different policy options.
- **Sectoral Analysis:** CGE models are useful for examining the impacts of climate policies on specific sectors. For example, they can analyze how changes in energy prices due to carbon pricing affect energy-intensive industries or how

agricultural policies aimed at reducing emissions influence food production and prices. This sectoral analysis helps policymakers design targeted interventions that minimize negative impacts and maximize positive outcomes.

Dynamic Stochastic General Equilibrium (DSGE) Models

Dynamic Stochastic General Equilibrium (DSGE) models extend the CGE framework by incorporating random shocks and dynamic adjustments. These models are particularly useful for analyzing the macroeconomic implications of climate risks and policy responses over time.

Structure and Functioning of DSGE Models

- **Dynamic Behaviour:** Unlike static CGE models, DSGE models account for the dynamic behaviour of economic agents. They use forward-looking equations to capture how households, firms, and the government make decisions over time, considering their expectations about future economic conditions.
- **Stochastic Elements:** DSGE models incorporate random shocks to represent unexpected events that can affect the economy. These shocks can be technological changes, policy shifts, or external factors such as climate events. By including stochastic elements, DSGE models provide a more realistic representation of economic dynamics and uncertainty.
- **Adjustment Mechanisms:** DSGE models include mechanisms for economic adjustment to shocks. For instance, they can simulate how firms adjust their production levels in response to changes in technology or how households alter their consumption patterns in response to income fluctuations. These adjustments help capture the short-term and long-term effects of shocks on the economy.

Applications of DSGE Models

- **Macroeconomic Analysis:** DSGE models are used for macroeconomic analysis, providing insights into how economies respond to various shocks and policy interventions over time. For example, they can analyze the effects of a sudden increase in carbon prices on GDP growth, inflation, and employment. By considering the dynamic adjustments of economic agents, DSGE models offer a comprehensive view of the macroeconomic impacts of climate policies.
- **Climate Risk Assessment:** In the context of climate risk assessment, DSGE models help evaluate the long-term implications of climate risks and adaptation strategies. For instance, they can simulate the economic impacts of increasing frequency and intensity of extreme weather events. These models can also assess the effectiveness of different policy responses, such as investing in climate-resilient infrastructure or implementing carbon pricing mechanisms.
- **Policy Optimization:** DSGE models are valuable for policy optimization, helping policymakers design strategies that balance economic growth and climate resilience. By simulating different policy scenarios, DSGE models can identify the optimal mix of mitigation and adaptation measures that minimize economic disruptions and maximize welfare. This optimization process considers the dynamic interactions between policies and the economy, ensuring that strategies are both effective and sustainable.

Conclusion

Both CGE and DSGE models are crucial for understanding the economic implications of climate risks and policy responses. CGE models provide detailed insights into the sectoral and overall economic impacts of policy changes and external shocks, making them valuable for policy analysis and climate risk management.

DSGE models extend this analysis by incorporating dynamic behaviour and stochastic elements, offering a more realistic representation of economic dynamics and uncertainty. Together, these models help policymakers design effective strategies to manage climate risks, promote economic resilience, and ensure sustainable development.

Cost-Benefit Analysis of Risk Management Strategies

Cost-Benefit Analysis (CBA) is a fundamental tool for evaluating the economic feasibility of climate risk management strategies.

By comparing the costs of implementing risk reduction measures with the anticipated benefits in terms of avoided damages and enhanced resilience, CBA helps policymakers determine the most efficient and effective ways to mitigate climate risks. This process involves a series of systematic steps to ensure that all relevant factors are considered and quantified.

Steps in Conducting CBA

- **Identifying Options:** The first step in conducting a CBA is to identify and list all potential risk management options. These options can be broadly categorized into structural measures, non-structural measures, and ecosystem-based approaches.
- **Structural Measures:** These include physical interventions designed to reduce vulnerability to climate hazards. Examples are building flood defences, reinforcing infrastructure, and constructing storm surge barriers. Structural measures often involve significant capital investment but can provide substantial protection against specific climate impacts.
- **Non-Structural Measures:** These encompass policy and regulatory changes aimed at enhancing resilience and reducing risks. Examples include land-use planning regulations, building codes, and early warning systems.

Non-structural measures are usually less capital-intensive than structural measures and can address a wide range of risks by influencing behaviour and decision-making processes.

- **Ecosystem-Based Approaches:** These strategies leverage natural systems to mitigate climate risks. Examples include wetland restoration, reforestation, and the creation of green infrastructure. Ecosystem-based approaches provide multiple co-benefits, such as enhancing biodiversity, improving water quality, and sequestering carbon, in addition to reducing climate risks.

Estimating Costs

Once the potential risk management options are identified, the next step is to estimate the costs associated with each option. This involves calculating both direct and indirect costs.

- **Direct Costs:** These are the immediate expenses related to the implementation of the risk management measure. They include capital investments for construction, operation and maintenance costs, and any required upgrades or replacements over time.
- **Indirect Costs:** These costs arise from the broader social and environmental impacts of the measures.

For instance, building a flood barrier may disrupt local ecosystems or displaced communities, leading to additional costs. It is important to capture these indirect costs to fully understand the economic implications of each option.

Estimating Benefits

Estimating the benefits of each risk management option involves quantifying the positive outcomes in terms of reduced damages, avoided losses, and other enhancements.

- **Reduced Damages:** This refers to the economic value of damages avoided due to the implementation of the risk management measure. For example, flood defences can prevent property damage, infrastructure losses, and business interruptions during flood events.

- **Avoided Losses:** These include economic losses that are prevented by the risk management measure. Examples are maintaining agricultural productivity during droughts, ensuring continuous operation of critical infrastructure, and preventing health-related costs associated with climate hazards.
- **Other Positive Outcomes:** Risk management measures can also lead to broader benefits such as improved public health, enhanced environmental quality, and increased economic resilience. For instance, ecosystem-based approaches like wetland restoration can enhance biodiversity, improve water quality, and provide recreational opportunities, all of which have positive economic value.

Discounting

Future costs and benefits must be discounted to their present values to account for the time value of money. This process involves applying a discount rate, which reflects the preference for immediate benefits over future benefits. Choosing an appropriate discount rate is critical, as it can significantly influence the outcome of the analysis.

- **Present Value Calculation:** The present value of future costs and benefits is calculated using the chosen discount rate. This allows for a fair comparison of costs and benefits that occur at different times. A lower discount rate generally increases the present value of future benefits, making long-term investments in risk management more attractive.
- **Comparing Costs and Benefits:** The net present value (NPV) of each risk management option is calculated by subtracting the present value of costs from the present value of benefits. Options with positive NPVs are considered economically viable, as their benefits outweigh their costs.

- **NPV Calculation:** The NPV provides a clear metric for comparing different options. A higher NPV indicates a more economically attractive option. Policymakers can use NPVs to prioritize risk management measures that offer the greatest net benefits to society.
- **Sensitivity Analysis:** Sensitivity analysis is conducted to test the robustness of the CBA results. This involves varying key assumptions and parameters to understand their impact on the outcome of the analysis.
- **Varying Assumptions:** Key assumptions such as discount rates, cost estimates, and benefit valuations are adjusted to see how sensitive the results are to changes in these variables. This helps identify the most critical factors influencing the economic viability of the risk management options.
- **Scenario Testing:** Different scenarios, such as varying levels of climate change severity, economic growth rates, and technological advancements, can be tested to assess how they affect the CBA outcomes. This ensures that the chosen risk management strategies are robust under a range of possible future conditions.

Conclusion

Cost-benefit analysis is an essential tool for evaluating the economic feasibility of climate risk management strategies. By systematically identifying options, estimating costs and benefits, discounting future values, comparing net present values, and conducting sensitivity analysis, policymakers can make informed decisions about the most effective and efficient ways to mitigate climate risks. Ensuring that these analyses are comprehensive and robust helps prioritize actions that provide the greatest net benefits to society, ultimately contributing to enhanced resilience and sustainable development in the face of climate change.

Integrating Climate Risk into Economic Planning

Integrating climate risk into economic planning is essential for building resilient economies and societies. This process involves embedding climate considerations into development policies, investment decisions, and financial planning.

By doing so, governments, businesses, and financial institutions can mitigate the adverse impacts of climate change and enhance economic resilience.

Strategies for Integration

Policy and Regulatory Frameworks

Governments play a pivotal role in integrating climate risks into economic planning by enacting policies and regulations that mandate the consideration of climate impacts in development projects. This involves incorporating climate risk assessments into land-use planning, infrastructure design, and environmental regulations.

- **Land-Use Planning:** Incorporating climate risk assessments into land-use planning helps ensure that new developments are located in areas less vulnerable to climate impacts, such as flooding, sea-level rise, or extreme weather events. Zoning regulations can restrict development in high-risk areas and promote the use of green spaces and natural buffers that enhance resilience.
- **Infrastructure Design:** Climate-resilient infrastructure design standards ensure that new infrastructure can withstand climate impacts. This includes building codes that require elevated structures in flood-prone areas, the use of durable materials that can endure extreme weather, and the incorporation of energy-efficient technologies to reduce greenhouse gas emissions.
- **Environmental Regulations:** Environmental regulations that consider climate risks can help protect natural ecosystems that provide essential services such as flood

protection, water purification, and carbon sequestration. Regulations can also promote sustainable land management practices that reduce vulnerability to climate change.

Public Investment and Budgeting

Integrating climate risk into public investment decisions is crucial for ensuring that infrastructure and public services are resilient to climate impacts. This involves allocating budgetary resources for adaptation measures and incorporating climate risk assessments into the appraisal of public investment projects.

- **Budget Allocation:** Governments can allocate funds specifically for climate adaptation projects, such as upgrading infrastructure to withstand extreme weather, enhancing water management systems, and developing early warning systems.
  By prioritizing these investments, governments can reduce the long-term economic costs of climate impacts.
- **Project Appraisal:** Incorporating climate risk assessments into the appraisal of public investment projects ensures that the potential impacts of climate change are considered during the planning stages. This helps avoid costly retrofits and repairs in the future and ensures that public investments are sustainable and resilient.

Private Sector Engagement

Encouraging the private sector to integrate climate risk into their business strategies and investment decisions is crucial. This can be achieved through incentives such as tax breaks for resilient investments, mandatory disclosure of climate risks, and support for climate-smart business practices.

- **Incentives:** Governments can provide tax incentives or subsidies for businesses that invest in climate resilience measures, such as energy-efficient buildings, renewable energy installations, and climate-resilient supply chains.

These incentives can reduce the upfront costs of adaptation and encourage widespread adoption of resilient practices.

- **Mandatory Disclosure:** Requiring businesses to disclose their exposure to climate risks and their strategies for managing these risks can increase transparency and accountability. This information helps investors make informed decisions and encourages companies to adopt more sustainable practices.
- **Climate-Smart Business Practices:** Supporting businesses in developing and implementing climate-smart practices can enhance their resilience and competitiveness. This includes providing technical assistance, access to financing, and platforms for knowledge exchange on best practices in climate risk management.

Financial Instruments

Developing and promoting financial instruments that support climate risk management can mobilize private and public capital for adaptation efforts. These instruments provide the financial resources needed to implement and sustain risk management measures.

- **Climate Bonds:** Climate bonds are a type of green bond specifically designed to fund projects that address climate change, such as renewable energy installations, energy efficiency upgrades, and climate-resilient infrastructure. By issuing climate bonds, governments and corporations can attract investment for adaptation projects.
- **Insurance Products:** Innovative insurance products can help manage financial risks associated with climate impacts. For example, parametric insurance pays out based on predefined triggers, such as rainfall levels or wind speeds, providing quick financial relief following extreme weather events. This helps businesses and communities recover more rapidly and reduce economic disruption.

- **Resilience Funds:** Establishing resilience funds can provide dedicated financial resources for climate adaptation projects. These funds can be used to support community-based adaptation initiatives, infrastructure upgrades, and other measures that enhance resilience to climate change.

Capacity Building

Strengthening the capacity of institutions and individuals to assess and manage climate risks is essential for effective integration. This includes training for policymakers, planners, and financial analysts on climate risk assessment methodologies and economic planning tools.

- **Training Programmes:** Comprehensive training programmes can equip government officials, business leaders, and financial professionals with the skills needed to integrate climate risk into their planning and decision-making processes. This includes training on using climate models, conducting vulnerability assessments, and evaluating the economic impacts of climate risks.
- **Knowledge Sharing:** Facilitating knowledge sharing and collaboration among different stakeholders can enhance capacity building efforts. This includes creating platforms for exchanging best practices, conducting joint research projects, and fostering partnerships between government agencies, businesses, and academic institutions.

Monitoring and Evaluation

Establishing systems for monitoring and evaluating the effectiveness of climate risk management strategies ensures that they are achieving their intended outcomes and allows for adjustments based on new information and changing conditions.

- **Performance Indicators:** Developing clear performance indicators helps track the progress of adaptation measures and assess their effectiveness. These indicators can include metrics such as reduced economic losses from climate impacts, improved infrastructure resilience, and enhanced community preparedness.

- **Adaptive Management:** Implementing an adaptive management approach allows for continuous learning and improvement. By regularly reviewing and updating climate risk management strategies based on monitoring and evaluation results, policymakers can ensure that measures remain relevant and effective in the face of evolving climate risks.

Conclusion

Integrating climate risk into economic planning is essential for building resilient economies and societies. This involves mainstreaming climate considerations into development policies, investment decisions, and financial planning. By enacting policy and regulatory frameworks, prioritizing public investment and budgeting, engaging the private sector, developing financial instruments, building capacity, and establishing robust monitoring and evaluation systems, governments and businesses can effectively manage climate risks. These efforts not only reduce the economic costs of climate impacts but also promote sustainable development and enhance overall resilience.

Effective climate risk assessment and management are critical for mitigating the economic impacts of climate change and enhancing resilience. By employing robust assessment methodologies, leveraging economic models, conducting thorough cost-benefit analyses, and integrating climate risk into economic planning, societies can develop comprehensive strategies that safeguard their economies and communities against climate-related risks. Ensuring that these efforts are inclusive and equitable is vital for addressing the disproportionate impacts of climate change on vulnerable populations. Through coordinated action and informed decision-making, it is possible to build a more resilient and sustainable future.

Comprehensive Climate Risk Assessment

A comprehensive approach to climate risk assessment begins with identifying and analyzing climate hazards such as extreme weather events, sea-level rise, and long-term temperature increases. Using historical data, climate models, and geographical information systems (GIS), this step maps out the frequency, intensity, and spatial distribution of these hazards. Following this, vulnerability assessments evaluate the susceptibility of different sectors, communities, and ecosystems. Factors such as socio-economic status, infrastructure robustness, and environmental conditions are considered to determine which areas and populations are most at risk.

Impact assessments then estimate the potential consequences of these climate hazards on economic activities, public health, infrastructure, and natural resources. By using climate models, economic models, and scenario analysis, this step provides a clearer picture of potential damages and disruptions, enabling more informed decision-making. Risk characterization integrates the findings from hazard, vulnerability, and impact assessments, quantifying the overall risk in terms of probability and magnitude. This comprehensive assessment forms the basis for prioritizing risk management actions and resources effectively.

Leveraging Economic Models for Risk Management

Economic models are indispensable tools in climate risk management, helping to analyze the costs and benefits of different strategies and integrate climate risks into economic planning. Cost-Benefit Analysis (CBA) is fundamental, comparing the costs of implementing risk reduction measures with the anticipated benefits from avoided damages and enhanced resilience. This analysis helps identify the most economically efficient strategies, ensuring that limited resources are used effectively. Integrated Assessment Models (IAMs) combine climate science, economics, and policy analysis to evaluate the interactions between climate change and economic systems.

These models are crucial for assessing the long-term economic impacts of climate change and the effectiveness of various mitigation and adaptation policies. Computable General Equilibrium (CGE) models simulate how economies respond to changes in policies, prices, and external shocks, providing insights into the broader economic effects of climate risk management strategies. Similarly, Dynamic Stochastic General Equilibrium (DSGE) models incorporate random shocks and dynamic adjustments, analysing the macroeconomic implications of climate risks and policy responses over time.

Conducting Thorough Cost-Benefit Analyses

Cost-Benefit Analysis (CBA) is a vital tool for evaluating the economic feasibility of climate risk management strategies. It involves several steps:

- **Identifying Options:** Listing all potential risk management options, including structural measures (e.g. building flood defences), non-structural measures (e.g. policy changes), and ecosystem-based approaches (e.g. wetland restoration).
- **Estimating Costs:** Calculating the direct costs of implementing each option, including capital investments, operation and maintenance costs, and potential indirect costs such as social and environmental impacts.
- **Estimating Benefits:** Quantifying benefits in terms of reduced damages, avoided losses, and other positive outcomes, such as economic value of avoided disruptions to economic activities, health benefits, and environmental improvements.
- **Discounting:** Discounting future costs and benefits to present values to account for the time value of money. The choice of discount rate is critical, as it significantly influences the analysis outcome.
- **Comparing Costs and Benefits:** Calculating the net present value (NPV) of each option by subtracting the present value of costs from the present value of benefits.

Options with positive NPVs are considered economically viable.

- **Sensitivity Analysis:** Testing the robustness of CBA results by varying key assumptions and parameters, such as discount rates, cost estimates, and benefit valuations.

Integrating Climate Risk into Economic Planning

Integrating climate risk into economic planning is essential for building resilient economies and societies. This involves mainstreaming climate considerations into development policies, investment decisions, and financial planning.

- **Policy and Regulatory Frameworks:** Governments can integrate climate risks by enacting policies and regulations that mandate the consideration of climate impacts in development projects. This includes incorporating climate risk assessments into land-use planning, infrastructure design, and environmental regulations.
- **Public Investment and Budgeting:** Integrating climate risk into public investment decisions ensure that infrastructure and public services are resilient to climate impacts. Allocating budgetary resources for adaptation measures and incorporating climate risk assessments into the appraisal of public investment projects are crucial steps.
- **Private Sector Engagement:** Encouraging the private sector to integrate climate risk into their business strategies and investment decisions is vital. This can be achieved through incentives such as tax breaks for resilient investments, mandatory disclosure of climate risks, and support for climate-smart business practices.
- **Financial Instruments:** Developing and promoting financial instruments that support climate risk management, such as climate bonds, insurance products, and resilience funds, can mobilize private and public capital for adaptation efforts.

- **Capacity Building:** Strengthening the capacity of institutions and individuals to assess and manage climate risks is essential. This includes training for policymakers, planners, and financial analysts on climate risk assessment methodologies and economic planning tools.
- **Monitoring and Evaluation:** Establishing systems for monitoring and evaluating the effectiveness of climate risk management strategies ensures they achieve their intended outcomes and allows for adjustments based on new information and changing conditions.

Ensuring Inclusivity and Equity

Addressing the disproportionate impacts of climate change on vulnerable populations is critical. Inclusive and equitable climate risk management ensures that the needs of marginalized communities are considered and addressed. Targeted interventions, capacity building, and support for these communities can enhance their resilience and contribute to broader social equity.

Coordinated Action and Informed Decision-making

Coordinated action and informed decision-making are key to effective climate risk management. Collaboration among governments, businesses, financial institutions, and civil society can enhance the effectiveness of adaptation strategies. Informed decision-making, supported by robust data and analysis, ensures that climate risk management efforts are efficient, effective, and responsive to changing conditions.

Building a Resilient and Sustainable Future

Integrating climate risk into economic planning, leveraging economic models, and conducting thorough cost-benefit analyses are essential steps in building resilient and sustainable economies. Ensuring that these efforts are inclusive and equitable addresses the disproportionate impacts of climate change on vulnerable populations. Through coordinated action and informed decision-making, societies can develop comprehensive strategies that safeguard their economies and communities against climate-related risks, paving the way for a more resilient and sustainable future.

REFERENCES

Aghion, P., Bergeaud, A., Cette, G. and Blundell, R. (2022). "The Impact of Inflation on Innovation: Evidence from Patent Data". *Journal of Monetary Economics, 131*: 77-92.

Bernanke, B.S. (2022). "The Real Effects of Disinflation". *Brookings Papers on Economic Activity, 2022*(1): 77-123.

Auerbach, A.J., and Gorodnichenko, Y. (2022). "Fiscal Multipliers in Recessions and Expansions: Does It Matter Whether Government Spending is Increasing or Decreasing?" *Journal of Economic Perspectives, 36*(1): 41-64.

Gali, J. (2022). *Monetary Policy, Inflation, and the Business Cycle: An Introduction to the New Keynesian Framework*, 3rd edn., Princeton University Press.

Gopinath, G. (2022). "The International Dimension of Inflation: Evidence and Policy Implications". *IMF Staff Papers, 69*(2): 189-218.

Gorton, G. and Metrick, A. (2022). "The Fed and the Great Inflation of the 1970s: The Evidence Speaks". *Journal of Monetary Economics, 124*: 94-114.

Guerrieri, V., Lorenzoni, G., Straub, L. and Werning, I. (2022). "Macroeconomic Implications of COVID-19: Can Negative Supply Shocks Cause Demand Shortages?" *American Economic Review: Insights, 4*(2): 173-190.

Ha, J., Kose, M.A. and Ohnsorge, F. (2022). "Global Inflation Synchronization". *Journal of International Economics, 134*, Editorial Board Article No. 103595.

Hall, R.E. (2022). "The Limits of Inflation Targeting". *Journal of Economic Perspectives, 36*(2): 3-24.

Jordà, Ò., Singh, S.R. and Taylor, A.M. (2022). "The Long-Run Effects of Monetary Policy". *Journal of Monetary Economics, 123*: 1-20.

Kamin, S.B. (2022). "Global Supply Chains in a Post-Pandemic World: Economic Insights and Policy Responses". *International Journal of Central Banking, 18*(3): 57-88.

Krugman, P. (2022). "Inflation and the Power of Economic Ideas". *Foreign Affairs, 101*(2): 23-37.

Lane, P.R. (2022). "Global Supply Chains and Inflation: The Role of the Pandemic and Policy Responses". *IMF Economic Review, 70*(2): 121-143.

Leduc, S. and Wilson, D. (2022). "What is Driving the Rise in Inflation? A Primer". *Economic Letter*, Federal Reserve Bank of San Francisco, 2022-27.

Martin, F. (2022). "Inflation, Wage Setting, and the Monetary Transmission Mechanism". *Journal of Economic Theory, 197*.

Mian, A., Sufi, A. and Verner, E. (2022). "Household Debt and Inflation: The Role of Mortgage Debt in Monetary Policy Transmission". *Journal of Financial Economics, 143*(3): 1041-1061.

Nakamura, E. and Steinsson, J. (2022). "Fiscal and Monetary Policy Coordination in a Low-Interest-Rate World". *Quarterly Journal of Economics, 137*(2): 617-660.

Obstfeld, M. (2022). "Global Macroeconomic Challenges: Inflation, Supply Chains, and Policy Responses". *Brookings Papers on Economic Activity, 2022*(2): 37-74.

Rogoff, K. (2022). "The Optimal Level of Inflation: Theory and Evidence". *Journal of Economic Perspectives, 36*(3): 67-90.

Shiller, R.J. (2022). *Narrative Economics: How Stories Go Viral and Drive Major Economic Events*, 2nd edn., Princeton University Press.

Summers, L.H. (2022). "Secular Stagnation and the Road to Recovery". *American Economic Review: Papers & Proceedings, 112*: 104-109.

Woodford, M. (2022). "Inflation and Financial Stability: A Microeconomic Perspective". *Journal of Economic Literature, 60*(2): 337-365.

4

Policy Frameworks for Economic Resilience in a Changing Climate

Key Policy Instruments for Economic Resilience

Climate Adaptation Policies

Climate adaptation policies are crucial for enhancing the capacity of societies to cope with the adverse effects of climate change. These policies aim to reduce vulnerability and increase resilience across various sectors through a wide range of strategies. One significant area of focus is infrastructure resilience, which involves setting building codes and standards to ensure new constructions can withstand extreme weather events, sea-level rise, and other climate impacts. For instance, regulations may require the elevation of buildings in flood-prone areas or the use of materials that can resist high winds and heat. Such measures help protect critical infrastructure and reduce the economic costs associated with climate-induced damages.

Ecosystem-Based Adaptation is another vital policy approach that leverages natural systems to mitigate climate risks. By promoting the restoration and conservation of ecosystems, these policies enhance the natural defences against climate impacts.

For example, restoring wetlands can absorb excess floodwaters, protecting surrounding communities from flooding. Similarly, conserving mangroves can shield coastlines from storm surges, and reforesting watersheds can stabilize soils and regulate water flow. These ecosystem-based strategies not only provide environmental benefits but also support biodiversity and improve the quality of life for local communities.

Disaster Risk Reduction (DRR) policies focus on preparing for and mitigating the impacts of natural disasters. These policies include the development and implementation of early warning systems, emergency response plans, and community-based preparedness programmes. Early warning systems, for instance, can provide timely information about impending disasters, allowing communities to take preventive measures. Emergency response plans ensure that resources and protocols are in place to efficiently manage the aftermath of disasters, minimizing economic disruptions and aiding in swift recovery.

Economic Incentives

Economic incentives are powerful tools for encouraging businesses and individuals to adopt practices that enhance climate resilience. "Subsidies and Grants" can be provided for climate-resilient technologies and practices, reducing the financial burden on those who invest in such measures. For instance, financial support for the installation of solar panels, energy-efficient appliances, or water-saving irrigation systems can promote sustainable practices and reduce vulnerability to climate impacts.Tax Incentives are another effective means of promoting resilience. Governments can offer tax credits or deductions to property owners who invest in flood-proofing their homes or to businesses that implement climate-smart agricultural practices.

These incentives lower the upfront costs of adaptation measures, making them more accessible and attractive. Innovative Insurance Products can also play a critical role in managing financial risks associated with climate impacts. Parametric insurance, for example, pays out based on predefined triggers such as rainfall levels or wind speeds, providing quick financial relief following extreme weather events. This helps businesses and communities recover more rapidly and reduce economic disruption. By offering tailored insurance solutions, insurers can encourage risk-aware behaviour and investments in resilience.

Public-Private Partnerships

Public-Private Partnerships (PPPs) are essential for mobilizing resources and expertise for climate resilience projects. By collaborating with the private sector, governments can leverage additional funding and technical skills. Resilience Bonds are an innovative financial instrument that can be used to fund large-scale infrastructure projects designed to withstand climate impacts. These bonds attract investment from private investors, who are repaid from the savings generated by reduced disaster recovery costs. Climate Resilience Funds can be established to pool resources from multiple sources, including international donors, national governments, and private investors. These funds can be used to finance a wide range of resilience-building activities, from community-based adaptation projects to large-scale infrastructure upgrades. By providing a dedicated source of funding, resilience funds ensure that resources are available to implement and sustain risk management measures.

Capacity Building and Education

Building the capacity of institutions and individuals to assess and manage climate risks is critical for effective adaptation. Training Programmes for policymakers, planners, and financial analysts on climate risk assessment methodologies and economic planning tools are essential. These programmes equip participants with the skills and knowledge needed to integrate climate considerations into their decision-making processes. Public Awareness Campaigns can educate communities about the risks of climate change and the importance of resilience. By raising awareness and fostering a culture of preparedness, these campaigns empower individuals to take proactive measures to protect themselves and their assets.

Monitoring and Evaluation

Establishing robust Monitoring and Evaluation (M&E) systems is vital for assessing the effectiveness of climate resilience policies. Performance Indicators should be developed to track progress and measure outcomes. These indicators can include metrics such as the reduction in economic losses from climate impacts, improvements in infrastructure resilience, and enhanced community preparedness. Regular Reporting and Reviews ensure accountability and transparency in resilience-building efforts. Governments should publish progress reports on their adaptation actions and conduct periodic reviews to assess the effectiveness of their policies and programmes. This continuous feedback loop allows for adjustments and improvements, ensuring that policies remain relevant and effective in the face of evolving climate risks.

Conclusion

Climate adaptation policies, economic incentives, public-private partnerships, capacity building, and robust monitoring and evaluation systems are key instruments for enhancing economic resilience to climate change. By integrating these elements into economic planning, governments and institutions can develop comprehensive strategies that protect economies and communities from the adverse impacts of climate change, ensuring a more resilient and sustainable future.

Ecosystem-based Adaptation

Ecosystem-based Adaptation (EbA) is a strategy that leverages natural systems to enhance resilience against climate impacts.

This approach recognizes that healthy ecosystems provide critical services that can buffer the effects of climate change, thereby reducing vulnerability and enhancing the adaptive capacity of communities. One of the key policies in EbA involves the "restoration of wetlands". Wetlands act as natural sponges, absorbing excess floodwaters during heavy rains and reducing the risk of flooding in adjacent areas. By restoring degraded wetlands, we can enhance their capacity to manage flood risks. Wetland restoration projects often involve replanting native vegetation, removing invasive species, and re-establishing natural water flow patterns.

These efforts not only mitigate flood risks but also provide habitat for wildlife, improve water quality, and support recreational activities, contributing to overall ecosystem health and community well-being. Another critical component of EbA is the "conservation of mangroves". Mangrove forests, found along coastlines in tropical and subtropical regions, provide a vital defence against storm surges and coastal erosion. Their dense root systems trap sediments and reduce wave energy, protecting shorelines from the destructive forces of storms and rising sea levels. Policies that support the conservation and sustainable management of mangroves include legal protections against deforestation, community-led conservation initiatives, and the integration of mangrove conservation into coastal development planning.

These measures ensure that mangroves continue to provide essential protective services while supporting local livelihoods through fisheries, tourism, and other economic activities. Reforestation of watersheds is another essential EbA strategy. Forests in watershed areas play a crucial role in stabilizing soils, regulating water flow, and maintaining water quality. By preventing soil erosion and reducing the risk of landslides, reforestation helps protect communities and infrastructure from the impacts of heavy rainfall and flooding. Reforestation projects typically involve planting native tree species, implementing sustainable forest management practices, and engaging local communities in forest conservation efforts. These projects enhance biodiversity, sequester carbon, and provide economic opportunities through sustainable forestry and eco-tourism.

Disaster Risk Reduction

Disaster Risk Reduction (DRR) policies focus on preparing for and mitigating the impacts of natural disasters. These policies are essential for enhancing the ability of populations to respond to and recover from disasters, reducing the economic and social costs associated with such events.

One of the foundational elements of DRR is the implementation of "early warning systems". Early warning systems provide timely and accurate information about impending disasters, allowing individuals and communities to take preventive measures. These systems involve monitoring hazards, forecasting potential impacts, and disseminating warnings through various communication channels, such as radio, television, mobile alerts, and community networks. Effective early warning systems require coordination between meteorological agencies, emergency services, and community organizations to ensure that warnings reach those at risk and that appropriate actions are taken. Emergency response plans are another critical aspect of DRR.

These plans outline the procedures and resources needed to respond to disasters effectively. They include strategies for evacuating vulnerable populations, providing emergency shelter and medical care, restoring essential services, and coordinating recovery efforts. Developing and regularly updating emergency response plans involves conducting risk assessments, identifying resource needs, training emergency personnel, and engaging with communities to ensure that everyone knows what to do in the event of a disaster. Community-based preparedness programmes play a vital role in enhancing resilience at the grassroots level. These programmes involve educating communities about disaster risks, promoting preparedness activities, and building local capacity to respond to emergencies. Community-based initiatives often include organizing drills and simulations, establishing local emergency response teams, and creating community disaster plans.

By involving community members in preparedness efforts, these programmes build a sense of ownership and responsibility, ensuring that people are better equipped to protect themselves and their neighbours during disasters. Moreover, integrating DRR into broader development planning is essential for sustainable risk management. This involves incorporating disaster risk assessments into land-use planning, infrastructure design, and environmental management. For instance, zoning regulations can prevent the construction of buildings in high-risk areas, and infrastructure projects can be designed to withstand extreme weather events. Additionally, environmental management practices that reduce disaster risks, such as maintaining healthy forests and wetlands, can be promoted through policy measures.

Conclusion

Ecosystem-based adaptation and disaster risk reduction are critical components of a comprehensive approach to building climate resilience. By leveraging natural systems through EbA, we can enhance the capacity of ecosystems to buffer climate impacts, providing multiple co-benefits for communities and the environment. Meanwhile, DRR policies ensure that populations are prepared for and can effectively respond to natural disasters, minimizing their impacts and facilitating swift recovery. Integrating these strategies into economic planning and development policies is essential for creating resilient societies capable of withstanding the challenges posed by climate change. Through coordinated efforts and informed decision-making, we can safeguard lives, livelihoods, and ecosystems, paving the way for a more sustainable and resilient future.

Economic Incentives

Economic incentives are powerful tools for encouraging businesses and individuals to adopt practices that enhance climate resilience. These incentives can take various forms, such as subsidies, grants, tax incentives, and financial instruments, all aimed at reducing the economic barriers to implementing climate-resilient measures.

Subsidies and Grants

Governments can provide subsidies and grants to promote the adoption of climate-resilient technologies and practices. For example, financial support for installing solar panels, energy-efficient appliances, or water-saving irrigation systems can significantly reduce vulnerability to climate impacts while promoting sustainability. Subsidies can lower the upfront costs of these technologies, making them more accessible to households and businesses. Grants can be targeted to specific projects, such as retrofitting buildings to withstand extreme weather or restoring ecosystems that provide natural flood protection. By reducing the financial burden, subsidies and grants can accelerate the transition to resilient and sustainable practices.

Tax Incentives

Tax incentives are another effective way to encourage climate-resilient investments. These incentives can include tax credits, deductions, and exemptions for activities that enhance resilience. For example, businesses that invest in renewable energy installations or adopt sustainable land management practices could receive tax credits, reducing their overall tax liability.

Home-owners who make energy-efficient upgrades to their properties could be eligible for tax deductions. Such incentives not only make resilient investments more financially attractive but also send a clear signal that sustainable practices are a priority for economic development. By aligning tax policies with resilience goals, governments can foster a culture of proactive climate adaptation.

Insurance Premium Reductions

Insurance companies can play a crucial role in promoting climate resilience through premium reductions for policyholders who implement risk-reducing measures. For instance, home-owners who install flood defences or fire-resistant materials could receive lower insurance premiums. Businesses that adopt comprehensive risk management plans could similarly benefit from reduced insurance costs. By offering financial incentives for risk mitigation, insurance companies encourage policyholders to take proactive steps to reduce their vulnerability to climate impacts. This not only benefits the individual policyholders but also reduces the overall risk exposure for insurance providers, creating a more resilient insurance market.

Climate Bonds

Climate bonds are financial instruments designed to raise capital for projects that have positive environmental benefits, including those that enhance climate resilience. Governments, development banks, and private companies can issue climate bonds to fund infrastructure projects, renewable energy installations, or ecosystem restoration efforts. Investors in these bonds are typically motivated by both financial returns and the desire to contribute to climate action. By providing a dedicated funding stream for resilience projects, climate bonds can mobilize substantial private sector investment, complementing public funding and accelerating the implementation of resilient measures.

Public-Private Partnerships

Public-Private Partnerships (PPPs) can be instrumental in leveraging private sector expertise and funding for climate resilience projects. In a PPP, the government collaborates with private companies to design, finance, and implement projects that enhance resilience. These partnerships can cover a wide range of activities, from building resilient infrastructure to developing early warning systems for natural disasters.

By sharing risks and benefits, PPPs can achieve more efficient and innovative solutions than either the public or private sector could accomplish alone. Additionally, PPPs can ensure that projects are aligned with both public policy goals and private sector interests, fostering a collaborative approach to resilience.

Carbon Pricing

Carbon pricing mechanisms, such as carbon taxes or cap-and-trade systems, can indirectly promote climate resilience by incentivizing the reduction of greenhouse gas emissions. By putting a price on carbon, these mechanisms encourage businesses to adopt cleaner technologies and more sustainable practices, which often coincide with resilience measures. For example, reducing reliance on fossil fuels and increasing energy efficiency can both lower emissions and enhance resilience to energy price volatility. Revenue generated from carbon pricing can also be reinvested into resilience projects, creating a virtuous cycle of sustainability and adaptation.

Conclusion

Economic incentives are essential for promoting the adoption of climate-resilient practices across various sectors. By reducing the financial barriers and providing tangible benefits for resilience investments, subsidies, grants, tax incentives, insurance premium reductions, climate bonds, public-private partnerships, and carbon pricing mechanisms can drive significant progress towards a more resilient future. These incentives not only enhance the capacity of communities and businesses to withstand climate impacts but also support broader economic development and environmental sustainability goals. Effective implementation of economic incentives requires coordinated efforts between governments, the private sector, and financial institutions to ensure that resilience becomes an integral part of economic planning and decision-making.

Tax Incentives

Tax incentives are powerful tools for encouraging investments in resilience-building measures by reducing the financial burden on property owners and businesses. These incentives can take the form of tax credits, deductions, and exemptions, each designed to promote activities that enhance climate resilience and reduce vulnerability to climate-related risks.

Tax Credits and Deductions

One of the most common forms of tax incentives is the tax credit, which directly reduces the amount of tax owed by an individual or business. For example, property owners who invest in flood-proofing their homes, such as elevating structures, installing sump pumps, or applying water-resistant materials, could receive a tax credit equivalent to a percentage of their investment.

This not only makes the initial expenditure more affordable but also encourages proactive measures to mitigate flood risks. Similarly, businesses that adopt climate-smart agricultural practices, such as using drought-resistant crops, implementing efficient irrigation systems, or engaging in soil conservation techniques, might qualify for tax deductions. These deductions reduce taxable income, thereby lowering the overall tax liability. By making these investments financially attractive, tax incentives can drive widespread adoption of practices that enhance resilience.

Tax Exemptions

Tax exemptions can also play a crucial role in promoting resilience investments. For instance, municipalities might offer property tax exemptions for buildings that meet certain resilience standards. This approach not only rewards property owners who invest in climate-proofing measures but also motivates others to follow suit, thereby increasing the overall resilience of the community. Additionally, sales tax exemptions on the purchase of climate-resilient materials and technologies can lower the cost of these essential items, making them more accessible to a broader audience.

Sector-Specific Incentives

Different sectors can benefit from tailored tax incentives that address their unique challenges and opportunities. For example, the agricultural sector could receive tax breaks for investments in sustainable farming practices, such as cover cropping, agro-forestry, and precision agriculture. These practices not only improve resilience to climate impacts like droughts and extreme weather but also enhance soil health and biodiversity. In the energy sector, tax incentives can promote the adoption of renewable energy sources and energy-efficient technologies.

For instance, tax credits for installing solar panels or wind turbines can accelerate the transition to clean energy, reducing greenhouse gas emissions and increasing energy security. Similarly, deductions for investments in energy-efficient building retrofits can lower operational costs and enhance resilience to power disruptions.

Long-term Benefits

The long-term benefits of tax incentives for resilience investments are substantial. By encouraging property owners and businesses to implement climate-proofing measures, these incentives can reduce the economic and social costs associated with climate-related disasters. Enhanced resilience means fewer disruptions to economic activities, lower repair and recovery costs, and improved public safety. Moreover, tax incentives can stimulate economic growth by creating demand for resilience-related products and services, thereby supporting job creation and innovation in the green economy.

Insurance Products

Innovative insurance products are another critical tool for managing climate risks and enhancing resilience. Traditional insurance policies often fall short in addressing the specific and growing challenges posed by climate change. However, new insurance products, such as parametric insurance, can provide more effective financial protection against climate-related losses.

Parametric Insurance

Parametric insurance is an innovative approach that pays out based on predefined triggers, such as specific climate thresholds, rather than on assessed damages. For example, a parametric insurance policy might provide coverage that activates when rainfall exceeds a certain amount or wind speeds surpass a specific threshold. This type of insurance offers several advantages over traditional insurance models.

- **Quick Payouts:** One of the primary benefits of parametric insurance is the speed of payouts. Since payments are triggered automatically when predefined conditions are met, policyholders receive funds quickly, often within days. This rapid response is crucial for businesses and communities that need immediate financial support to recover from climate-related events, such as floods, hurricanes, or droughts.
- **Simplicity and Transparency:** Parametric insurance policies are straightforward and transparent. Policyholders know exactly what triggers a payout and the amount they will receive, eliminating the complexities and delays associated with damage assessments and claims processing. This clarity helps businesses and communities plan and manage their risks more effectively.
- **Coverage for Hard-to-Ensure Risks:** Traditional insurance products often excludes certain climate risks or provide inadequate coverage. Parametric insurance can fill these gaps by offering protection for hard-to-insure risks, such as agricultural losses due to extreme weather or infrastructure damage from sea-level rise. By addressing these vulnerabilities, parametric insurance enhances overall resilience.

Micro Insurance

Micro insurance is another innovative product designed to provide affordable coverage to low-income individuals and small businesses that are particularly vulnerable to climate impacts. These policies typically offer lower premiums and coverage amounts, making them accessible to those with limited financial resources.

- **Community-Based Insurance:** Community-based micro insurance programmes can pool resources from multiple members of a community to provide coverage against common risks. For example, a cooperative of small farmers might collectively purchase micro insurance to protect against crop failures due to drought. By sharing the risk, these programmes make insurance more affordable and accessible, enhancing the resilience of vulnerable populations.

- **Index-based Insurance:** Similar to parametric insurance, index-based micro insurance policies provide payouts based on specific indices, such as rainfall levels or temperature changes. This approach simplifies the claims process and ensures timely payouts, helping policyholders recover quickly from climate-related losses. Index-based insurance is particularly effective in agricultural settings, where timely financial support is critical for maintaining livelihoods.

Government and Donor

Support Government and donor support are essential for the development and scaling of innovative insurance products. Public-private partnerships can facilitate the design and implementation of parametric and micro insurance programmes, leveraging the expertise and resources of both sectors.

Additionally, subsidies and financial assistance from governments and international donors can reduce premiums and make these insurance products more accessible to vulnerable populations.

Conclusion

Economic incentives, such as tax incentives and innovative insurance products, are vital for promoting investments in climate resilience. Tax incentives, including credits, deductions, and exemptions, can reduce the financial burden on property owners and businesses, encouraging them to adopt resilience-building measures. These incentives can be tailored to address the specific needs of different sectors, driving widespread adoption of practices that enhance resilience and sustainability. Innovative insurance products, such as parametric insurance and micro insurance, provide effective financial protection against climate risks.

By offering quick payouts, simplicity, and coverage for hard-to-insure risks, these products help businesses and communities recover rapidly from climate-related losses. Government and donor support are crucial for developing and scaling these insurance products, ensuring they are accessible to those who need them most. Together, tax incentives and innovative insurance products can create a comprehensive framework for enhancing climate resilience, reducing vulnerability, and promoting sustainable economic development. By leveraging these tools, societies can build a more resilient and secure future in the face of growing climate challenges.

International Agreements and National Policies

International Agreements

International agreements play a crucial role in fostering global cooperation and commitment to climate resilience. By setting common goals and frameworks, these agreements encourage countries to work together to address the global challenge of climate change.

The Paris Agreement

The Paris Agreement, adopted under the United Nations Framework Convention on Climate Change (UNFCCC) in 2015, is a landmark international accord aimed at limiting global warming to well below 2°C above pre-industrial levels, with efforts to limit the increase to 1.5°C. The agreement recognizes the need for a robust and collective response to climate change, integrating both mitigation and adaptation efforts.

One of the critical aspects of the Paris Agreement is its emphasis on adaptation alongside mitigation. The agreement calls for enhanced support for developing countries in building resilience to climate impacts. This includes financial assistance, technology transfer, and capacity building to help these countries implement effective adaptation strategies. Developed countries are urged to provide financial resources to support adaptation and resilience-building activities in vulnerable nations. The Green Climate Fund (GCF), established under the UNFCCC, plays a significant role in channelling these resources to where they are most needed.

The Paris Agreement also introduces the concept of Nationally Determined Contributions (NDCs), where each country outlines its climate action plan, including adaptation measures. Countries are encouraged to integrate climate resilience into their national planning processes and to report on their progress regularly. This transparency and accountability mechanism helps ensure that countries are on track to meet their climate goals and enhances global cooperation.

Sendai Framework for Disaster Risk Reduction

The Sendai Framework for Disaster Risk Reduction, adopted in 2015, outlines global priorities for reducing disaster risk and enhancing resilience. It emphasizes the need for comprehensive risk assessments, inclusive and equitable policies, and international cooperation in disaster preparedness, response, and recovery.

The framework sets out four priorities for action:

- **Understanding Disaster Risk:** This priority emphasizes the importance of having a comprehensive understanding of disaster risk in all its dimensions, including exposure, vulnerability, and hazard characteristics. It calls for investment in data collection, risk assessments, and the dissemination of risk information to all stakeholders.
- **Strengthening Disaster Risk Governance:** Effective disaster risk governance involves coordination and collaboration among various stakeholders, including governments, the private sector, civil society, and international organizations. The Sendai Framework advocates for integrated and inclusive approaches to disaster risk management, ensuring that policies are

coherent and well-coordinated across different sectors and levels of government.

- **Investing in Disaster Risk Reduction for Resilience:** This priority highlights the need for investments in disaster risk reduction (DRR) measures that enhance resilience. This includes structural measures, such as resilient infrastructure and buildings, as well as non-structural measures, such as policies, regulations, and community-based approaches. The framework encourages the integration of DRR into national and local development planning and budgeting processes.
- **Enhancing Disaster Preparedness for Effective Response** and to "Build Back Better" in Recovery, Rehabilitation, and Reconstruction—Effective disaster preparedness involves having well-developed emergency plans, early warning systems, and response capacities. The Sendai Framework promotes the concept of "building back better" in recovery, rehabilitation, and reconstruction efforts, ensuring that communities are more resilient to future disasters.

National Policies

At the national level, governments play a critical role in implementing policies that promote climate resilience and disaster risk reduction. These policies are essential for translating international commitments into concrete actions that protect communities and economies from the adverse impacts of climate change.

- **National Adaptation Plans (NAPs):** National Adaptation Plans (NAPs) are a key instrument for integrating climate resilience into national development planning. NAPs provide a structured process for identifying and prioritizing adaptation needs, developing strategies, and mobilizing resources for implementation. They involve comprehensive assessments of climate vulnerabilities, stakeholder consultations, and the formulation of sector-specific adaptation actions. For example, a country with a significant agricultural sector might prioritize climate-smart agricultural practices in its NAP, including drought-resistant crops, efficient irrigation systems, and soil conservation techniques. Similarly, a coastal nation might focus on protecting critical infrastructure from sea-level rise and storm surges through the construction of seawalls, the restoration of mangroves, and the implementation of early warning systems.

- **Climate Resilience Strategies:** National governments also develop broader climate resilience strategies that encompass various sectors and scales. These strategies often include measures to enhance the resilience of infrastructure, such as roads, bridges, and buildings, to climate impacts. They also address the resilience of natural systems, such as watersheds and forests, which provide critical ecosystem services.
- **Disaster Risk Reduction Policies:** Disaster risk reduction (DRR) policies are essential for managing the risks associated with natural hazards. These policies include regulations, standards, and practices designed to reduce vulnerability and enhance preparedness. For instance, building codes that require structures to withstand extreme weather events, land-use planning that avoids high-risk areas, and emergency response plans that ensure efficient and coordinated action during disasters are all components of effective DRR policies.
- **Public Investment and Budgeting:** Integrating climate risk into public investment and budgeting processes is crucial for ensuring that resilience-building measures are adequately funded and prioritized. Governments allocate resources for climate adaptation projects, such as upgrading infrastructure, improving water management systems, and developing early warning systems. By incorporating climate risk assessments into public investment decisions, governments can avoid costly retrofits and repairs in the

future and ensure that public investments are sustainable and resilient.

- **Private Sector Engagement:** Engaging the private sector in climate resilience efforts is vital for leveraging additional resources and expertise. Governments can create an enabling environment for private sector investment in resilience by providing incentives, such as tax breaks, subsidies, and regulatory support. Public-Private Partnerships (PPPs) can also be instrumental in financing and implementing resilience projects, combining public sector oversight with private sector innovation and efficiency.
- **Monitoring and Evaluation:** Monitoring and Evaluation (M&E) systems are essential for assessing the effectiveness of climate resilience policies and programmes. By developing clear indicators and metrics, governments can track progress, measure outcomes, and identify areas for improvement.

Regular reporting and reviews ensure accountability and transparency, allowing for adjustments based on new information and changing conditions.

Conclusion

International agreements and national policies are fundamental to fostering global cooperation and commitment to climate resilience. The Paris Agreement and the Sendai Framework for Disaster Risk Reduction provide overarching frameworks for countries to develop and implement effective resilience strategies. At the national level, governments play a critical role in translating these international commitments into concrete actions through policies, regulations, and investment decisions. By integrating climate risk into development planning, engaging the private sector, and establishing robust monitoring and evaluation systems, countries can enhance their resilience to climate impacts and reduce the economic and social costs associated with disasters. Through coordinated efforts and informed decision-making, it is possible to build a more resilient and sustainable future.

National Adaptation Plans (NAPs)

Introduction

National Adaptation Plans (NAPs) have become a cornerstone of climate resilience strategies worldwide. These plans offer a systematic approach for countries to assess their climate vulnerabilities, prioritize adaptation actions, and integrate climate risks into national development planning. By providing a structured framework, NAPs enable countries to mobilize resources, coordinate efforts across sectors, and ensure that adaptation measures are sustainable and effective.

Framework for Integration

NAPs provide a comprehensive framework for integrating climate risk into national development planning. This integration is critical for ensuring that climate adaptation is not treated as a standalone issue but is embedded within broader socio-economic and development goals. This holistic approach helps countries to build resilience across various sectors, including agriculture, water resources, infrastructure, and public health. For example, in the agriculture sector, a NAP might include measures such as promoting climate-smart agricultural practices, developing drought-resistant crop varieties, and improving irrigation efficiency.

These measures not only reduce vulnerability to climate impacts but also enhance food security and livelihoods. In the water sector, NAPs might focus on improving water management, enhancing water storage capacity, and protecting water sources from contamination due to climate change.

Prioritizing Adaptation Actions

One of the primary functions of NAPs is to prioritize adaptation actions based on a comprehensive assessment of climate vulnerabilities. This prioritization ensures that limited resources are directed towards the most critical and impactful measures. The process typically involves extensive stakeholder consultations, vulnerability assessments, and cost-benefit analyses. Stakeholder consultations are essential for understanding local needs, capacities, and perspectives. Engaging communities, civil society organizations, and the private sector in the planning process helps to ensure that adaptation actions are context-specific and address the actual challenges faced by vulnerable populations.

Vulnerability assessments, on the other hand, provide a scientific basis for identifying areas and sectors that are most at risk from climate impacts. These assessments often involve the use of climate models, socio-economic data, and geographic information systems (GIS) to map vulnerabilities and project future risks. Cost-benefit analyses are used to evaluate the economic feasibility of different adaptation options. By comparing the costs of implementation with the anticipated benefits, such as reduced damages and enhanced resilience, countries can identify the most cost-effective strategies. This approach helps to maximize the return on investment and ensures that adaptation actions provide tangible benefits to communities and economies.

Mobilizing Resources for Implementation

Implementing NAPs requires significant financial, technical, and human resources. Mobilizing these resources is a critical challenge, especially for developing countries with limited fiscal capacity. NAPs provide a framework for identifying funding needs, exploring financing options, and coordinating resource mobilization efforts. International climate finance mechanisms, such as the Green Climate Fund (GCF) and the Global Environment Facility (GEF), play a crucial role in supporting the implementation of NAPs. These mechanisms provide grants, loans, and technical assistance to help countries develop and execute their adaptation plans.

Additionally, bilateral and multilateral donors, international development banks, and private sector investments contribute to financing adaptation projects. At the national level, governments can allocate budgetary resources for adaptation through public investment programmes and budgetary planning. Integrating climate risk into public investment decisions ensures that infrastructure and public services are resilient to climate impacts. For example, governments might allocate funds for upgrading flood defences, improving water management systems, and developing early warning systems.

- **Coordinating Efforts Across Sectors:** Climate adaptation requires a coordinated effort across multiple sectors and levels of government. NAPs facilitate this coordination by providing a common framework for action and promoting inter-sectoral collaboration. By aligning adaptation efforts with national development goals, NAPs help to ensure that adaptation measures are integrated into sectoral policies, plans, and programmes. For instance, in the health sector, a NAP might include measures to strengthen health systems' capacity to respond to climate-related health risks, such as heat waves, vector-borne diseases, and food and water-borne illnesses. In the infrastructure sector, NAPs might focus on enhancing the resilience of critical infrastructure, such as roads, bridges, and buildings, to climate impacts. By coordinating efforts across sectors, NAPs help to create synergies, avoid duplication, and maximize the effectiveness of adaptation actions.
- **Ensuring Sustainability and Effectiveness:** Ensuring the sustainability and effectiveness of adaptation measures is a key objective of NAPs. This requires continuous monitoring, evaluation, and learning to track progress, assess outcomes, and make necessary adjustments. NAPs include provisions for establishing monitoring and evaluation (M&E) systems that use clear indicators and metrics to measure the effectiveness of adaptation actions. Regular reporting and reviews help to ensure accountability and transparency. By tracking progress and evaluating outcomes, countries can identify successful strategies,

scale up best practices, and address any gaps or challenges. This iterative process allows for adaptive management, where adaptation measures are continuously refined and improved based on new information and changing conditions.

Conclusion

National Adaptation Plans (NAPs) are a vital tool for building climate resilience at the national level.

By providing a structured framework for integrating climate risk into development planning, prioritizing adaptation actions, mobilizing resources, coordinating efforts across sectors, and ensuring sustainability, NAPs help countries to systematically address their climate vulnerabilities and enhance their adaptive capacity. Through comprehensive planning, stakeholder engagement, and continuous learning, NAPs enable countries to develop and implement effective adaptation strategies that protect their economies, communities, and ecosystems from the adverse impacts of climate change. As the climate continues to change, the importance of NAPs will only grow, making them a critical component of global efforts to build a more resilient and sustainable future.

The Role of Governments and Institutions

Government Leadership

Governments at all levels play a critical role in driving climate resilience through policy development, implementation, and enforcement. Their leadership is essential for creating an enabling environment that supports and sustains adaptation efforts across sectors and communities.

Policy Development

Governments are responsible for developing policies that address climate risks and promote resilience. This process begins with conducting comprehensive vulnerability assessments to identify the most at risk areas and sectors. These assessments involve gathering and analyzing data on climate hazards, exposure, and socio-economic factors that contribute to vulnerability. By understanding these dynamics, governments can design targeted adaptation strategies that effectively address the specific needs of their populations. Stakeholder consultation is a crucial aspect of policy development.

Engaging with communities, businesses, non-governmental organizations (NGOs), and other stakeholders ensures that the policies are inclusive and consider the diverse perspectives and needs of different groups. This collaborative approach helps to build broad-based support for the policies and increases their chances of successful implementation. Comprehensive adaptation strategies are then formulated, outlining specific actions, timelines, and responsibilities. These strategies often cover multiple sectors, such as agriculture, water resources, infrastructure, health, and are tailored to national and local contexts.

For example, a national adaptation strategy might include measures to improve agricultural resilience through climate-smart farming practices, enhance water management systems to cope with changing rainfall patterns, and upgrade infrastructure to withstand extreme weather events.

Regulation and Enforcement

Effective regulation and enforcement are crucial for ensuring that resilience-building measures are implemented and maintained. Governments must establish clear regulations, such as building codes and land-use planning guidelines, that mandate the incorporation of climate resilience into development projects. Building codes can require structures to be designed and constructed to withstand specific climate hazards, such as hurricanes, floods, or heatwaves.

Land-use planning guidelines can restrict development in high-risk areas, such as floodplains or coastal zones, and promote the use of green infrastructure that enhances resilience. Enforcement mechanisms are necessary to ensure compliance with these regulations. This involves regular monitoring and inspections to verify that buildings and infrastructure meet the required standards. Non-compliance must be addressed through penalties or corrective actions to maintain the integrity of resilience measures. Effective enforcement also requires adequate resources and capacity within government agencies to carry out these functions.

Resource Allocation

Governments allocate resources for climate resilience, including budgetary allocations for infrastructure projects, research and development, and capacity-building initiatives. Strategic investment in resilience can reduce long-term economic costs and enhance overall societal well-being. Investing in resilient infrastructure is a key priority. This includes upgrading existing infrastructure to withstand climate impacts and constructing new infrastructure with resilience in mind. For instance, roads, bridges, and buildings can be designed to endure extreme weather events, while water supply and drainage systems can be improved to manage floods and droughts. Research and development (R&D) are essential for advancing climate resilience technologies and practices. Governments can fund R&D projects that focus on developing innovative solutions, such as drought-resistant crops, efficient irrigation systems, and renewable energy technologies.

These innovations can enhance adaptive capacity and reduce vulnerability to climate change. Capacity-building initiatives aim to strengthen the ability of individuals, communities, and institutions to assess and manage climate risks. This includes training programmes for government officials, planners, and financial analysts on climate risk assessment methodologies and economic planning tools. By building local capacity, governments can ensure that adaptation measures are effectively implemented and sustained over the long-term.

Institutional Support

Institutions, including international organizations, non-governmental organizations (NGOs), and the private sector, support government efforts by providing technical expertise, funding, and implementation support.

International Organizations

International organizations play a crucial role in supporting national and local governments in their climate resilience efforts. Key organizations, such as the World Bank, the United Nations Development Programme (UNDP), and the Global Environment Facility (GEF), provide both financial and technical assistance for climate resilience projects. The World Bank, for instance, funds large-scale infrastructure projects aimed at enhancing climate resilience, such as flood control systems, resilient housing, and sustainable urban planning initiatives. The Bank also provides technical assistance to help countries design and implement these projects effectively, ensuring that they are grounded in the latest scientific knowledge and best practices.

The UNDP focuses on capacity building and knowledge sharing. It works with countries to develop national adaptation plans (NAPs), conduct vulnerability assessments, and implement community-based adaptation projects. By providing training and resources, the UNDP helps countries strengthen their institutional capacity to manage climate risks. The GEF, through its various funding mechanisms, supports a wide range of climate resilience projects.

It provides grants and concessional loans to finance adaptation initiatives, particularly in developing countries. The GEF also facilitates the dissemination of best practices and innovative approaches to climate resilience, helping countries learn from each other's experiences and successes.

NGOs and Civil Society

Non-Governmental Organizations (NGOs) and civil society organizations are pivotal in advocating for climate resilience, raising awareness, and supporting community-based adaptation initiatives. They often serve as intermediaries between governments and local communities, ensuring that adaptation policies and programmes are inclusive and equitable. NGOs such as the International Red Cross and Red Crescent Societies, CARE International, and the World Wildlife Fund (WWF) implement a wide range of projects aimed at enhancing community resilience. These projects include building early warning systems, promoting sustainable agricultural practices, and restoring ecosystems to provide natural defences against climate impacts.

One of the key strengths of NGOs is their ability to work directly with communities. They engage local populations in the planning and implementation of adaptation measures, ensuring that these measures are culturally appropriate and address the specific needs and vulnerabilities of the community. This grassroots approach not only enhances the effectiveness of adaptation initiatives but also empowers communities to take ownership of their resilience strategies. Civil society organizations also play a crucial role in advocacy and education. They raise public awareness about climate risks and the importance of resilience, mobilize community action, and advocate for policy changes at local, national, and international levels. By amplifying the voices of vulnerable populations, these organizations help ensure that adaptation efforts are inclusive and socially just.

Private Sector

The private sector is a key driver of innovation and investment in resilience-building technologies and practices. Companies can incorporate climate risk into their business strategies, invest in resilient infrastructure, and develop new products and services that enhance resilience. For instance, insurance companies are developing innovative products to help businesses and communities manage climate risks. Parametric insurance policies, which pay out based on predefined triggers such as specific weather events, provide quick and efficient financial relief after disasters, enabling faster recovery.

Technology companies are also at the forefront of developing solutions for climate resilience. Advances in data analytics, artificial intelligence, and remote sensing are improving our ability to predict and respond to climate impacts. These technologies enable more accurate climate modelling, real-time monitoring of environmental conditions, and efficient management of resources. Moreover, the construction and engineering sectors are investing in resilient infrastructure. Companies are designing buildings and infrastructure that can withstand extreme weather events, incorporating sustainable materials, and adopting green building practices.

These investments not only reduce vulnerability to climate impacts but also promote long-term sustainability. Public-Private Partnerships (PPPs) are particularly effective in financing and implementing large-scale resilience projects. By combining the resources and expertise of both the public and private sectors, PPPs can achieve more efficient and innovative solutions to climate challenges. For example, a PPP might involve a private company constructing a resilient water supply system while the government provides regulatory support and ensures that the project aligns with national adaptation goals.

Conclusion

Institutional support from international organizations, NGOs, civil society, and the private sector is essential for advancing climate resilience efforts. International organizations provide critical financial and technical assistance, helping countries develop and implement effective adaptation strategies. NGOs and civil society organizations advocate for inclusive and equitable adaptation measures, support community-based initiatives, and raise public awareness. The private sector drives innovation and investment in resilience technologies and practices, contributing to sustainable development. Through collaborative efforts and shared expertise, these institutions play a vital role in building a more resilient and climate-adaptive world.

Evaluating Policy Effectiveness

Evaluating the effectiveness of climate resilience policies is crucial for ensuring that adaptation measures achieve their intended outcomes and provide lasting benefits. A robust framework for Monitoring and Evaluation (M&E), stakeholder engagement, and adaptive management is essential for assessing progress, identifying areas for improvement, and making informed adjustments to policies and programmes.

Monitoring and Evaluation Frameworks

- **Indicators and Metrics:** Developing clear indicators and metrics for resilience is critical for tracking progress and measuring outcomes. These indicators should be specific, measurable, achievable, relevant, and time-bound (SMART). For example, indicators might include the reduction in economic losses from climate impacts, improvements in infrastructure resilience, and enhanced community preparedness. Quantitative metrics such as the number of households protected from flooding, the extent of rehabilitated wetlands, or the percentage increase in crop yields due to climate-smart agricultural practices can provide concrete evidence of progress. Qualitative metrics, such as community perceptions of safety and resilience, are also important for capturing the broader impacts of adaptation measures.
- **Data Collection and Analysis:** Effective M&E requires robust data collection and analysis systems. Governments and institutions must invest in data infrastructure that allows for the systematic gathering, storage, and analysis of climate risk and resilience data. This includes geographic

information systems (GIS), remote sensing technologies, and databases for tracking project outcomes and climate impacts. Reliable data is essential for making evidence-based decisions and for evaluating the effectiveness of resilience-building measures.

- **Regular Reporting and Reviews:** Regular reporting and reviews ensure accountability and transparency in resilience-building efforts. Governments should publish progress reports on their adaptation actions and conduct periodic reviews to assess the effectiveness of their policies and programmes. These reports should detail the achievements, challenges, and lessons learned from implementing adaptation measures. Regular reviews allow for the identification of gaps and areas for improvement, ensuring that adaptation strategies remain relevant and effective.

Stakeholder Engagement

- **Community Involvement:** Involving local communities in the evaluation process ensures that their experiences and perspectives are considered. Community feedback can provide valuable insights into the effectiveness of adaptation measures and highlight areas for improvement. Participatory evaluation methods, such as community workshops, focus groups, and surveys, can be used to gather input from local residents. This engagement fosters a sense of ownership and accountability, as communities are directly involved in assessing the success of resilience initiatives.
- **Multi-Sectoral Collaboration:** Collaboration across sectors, including government agencies, the private sector, and civil society, ensures a comprehensive approach to evaluation. Multi-sectoral collaboration allows for the sharing of expertise, resources, and best practices. For instance, government agencies can provide regulatory oversight and funding, while the private sector can

contribute technical expertise and innovation. Civil society organizations can offer on-the-ground insights and advocate for the needs of vulnerable populations. This collaborative approach enhances the effectiveness and sustainability of resilience-building efforts.

Adaptive Management

- **Learning from Experience:** Adaptive management is a dynamic approach that involves learning from the successes and failures of implemented policies and programmes. By analysing what works and what does not, policymakers can refine their strategies and enhance their effectiveness. This iterative process allows for continuous improvement based on real-world experiences. For example, if a particular flood protection measure proves effective in one region, it can be scaled up and replicated in other areas. Conversely, if a strategy fails to deliver the expected results, it can be modified or replaced with a more effective approach.
- **Flexibility and Adjustments:** Climate resilience policies must be flexible and adaptable to changing conditions. Adaptive management allows for adjustments based on new information, emerging risks, and evolving societal needs. This flexibility is crucial for responding to the dynamic nature of climate change, which can present new challenges and opportunities over time. For instance, as climate models improve and provide more accurate projections, policies can be updated to reflect the latest scientific understanding. Similarly, as new technologies and practices are developed, they can be incorporated into resilience strategies.

Conclusion

Evaluating the effectiveness of climate resilience policies is essential for ensuring that adaptation measures provide tangible benefits and contribute to long-term sustainability. A robust M&E framework, encompassing clear indicators and metrics, comprehensive data collection and analysis, and regular reporting and reviews, is vital for tracking progress and measuring outcomes. Engaging stakeholders, including local communities and multi-sectoral collaborators, enhances the relevance and inclusiveness of evaluation processes.

Adaptive management allows for continuous learning and improvement, ensuring that policies remain flexible and responsive to changing conditions. Through these comprehensive evaluation efforts, policymakers can develop and implement more effective climate resilience strategies, safeguarding communities and economies against the adverse impacts of climate change. Integrating climate risk into economic planning is essential for building resilient economies and societies. Key policy instruments, including climate adaptation policies, economic incentives, and innovative financial products, play a crucial role in enhancing resilience.

International agreements and national policies provide a framework for global cooperation and commitment to climate resilience. Governments and institutions are pivotal in driving and supporting resilience efforts through policy development, regulation, resource allocation, and technical assistance. Evaluating the effectiveness of resilience policies through robust M&E frameworks, stakeholder engagement, and adaptive management ensures continuous improvement and sustained progress. By adopting comprehensive and inclusive approaches to climate risk management, societies can safeguard their economies and communities against the adverse impacts of climate change, paving the way for a more resilient and sustainable future.

REFERENCES

Aghion, P., Bergeaud, A., Cette, G. and Blundell, R. (2022). "The Impact of Inflation on Innovation: Evidence from Patent Data". *Journal of Monetary Economics, 131*: 77-92.

Bernanke, B.S. (2022). "The Real Effects of Disinflation". *Brookings Papers on Economic Activity, 2022*(1): 77-123.

Ha, J., Kose, M.A. and Ohnsorge, F. (2022). "Global Inflation Synchronization". *Journal of International Economics, 134*, Editorial Board Article No. 103595.

Hall, R.E. (2022). "The Limits of Inflation Targeting". *Journal of Economic Perspectives, 36*(2): 3-24.

Jordà, Ò., Singh, S.R. and Taylor, A.M. (2022). "The Long-Run Effects of Monetary Policy". *Journal of Monetary Economics, 123*: 1-20.

Kamin, S.B. (2022). "Global Supply Chains in a Post-Pandemic World: Economic Insights and Policy Responses". *International Journal of Central Banking, 18*(3): 57-88.

Krugman, P. (2022). "Inflation and the Power of Economic Ideas". *Foreign Affairs, 101*(2): 23-37.

Lane, P.R. (2022). "Global Supply Chains and Inflation: The Role of the Pandemic and Policy Responses". *IMF Economic Review, 70*(2): 121-143.

Leduc, S. and Wilson, D. (2022). "What is Driving the Rise in Inflation? A Primer". *Economic Letter*, Federal Reserve Bank of San Francisco, 2022-27.

Mian, A., Sufi, A. and Verner, E. (2022). "Household Debt and Inflation: The Role of Mortgage Debt in Monetary Policy Transmission". *Journal of Financial Economics, 143*(3): 1041-1061.

Nakamura, E. and Steinsson, J. (2022). "Fiscal and Monetary Policy Coordination in a Low-Interest-Rate World". *Quarterly Journal of Economics, 137*(2): 617-660.

Obstfeld, M. (2022). "Global Macroeconomic Challenges: Inflation, Supply Chains, and Policy Responses". *Brookings Papers on Economic Activity, 2022*(2): 37-74.

Rogoff, K. (2022). "The Optimal Level of Inflation: Theory and Evidence". *Journal of Economic Perspectives, 36*(3): 67-90.

Shiller, R.J. (2022). "Narrative Economics: How Stories Go Viral and Drive Major Economic Events", 2nd edn., Princeton University Press.

Summers, L.H. (2022). "Secular Stagnation and the Road to Recovery". *American Economic Review: Papers & Proceedings, 112*: 104-109.

Woodford, M. (2022). "Inflation and Financial Stability: A Microeconomic Perspective". *Journal of Economic Literature, 60*(2): 337-365.

5

Investing in Climate-Resilient Infrastructure and Innovation

Importance of Climate-Resilient Infrastructure

Climate-resilient infrastructure is critical for mitigating the adverse impacts of climate change and ensuring the long-term sustainability and safety of communities. As climate change continues to exacerbate the frequency and intensity of extreme weather events, such as floods, hurricanes, droughts, and heatwaves, the vulnerability of existing infrastructure becomes increasingly apparent. Traditional infrastructure, designed based on historical climate patterns, is often ill-equipped to handle these new extremes. Investing in climate-resilient infrastructure helps to safeguard essential services, protect lives and property, and reduce economic losses.

For instance, resilient infrastructure can prevent significant damage during extreme weather events, thereby avoiding costly repairs and disruptions to economic activities. More importantly, it can save lives by providing safe and secure environments during disasters. Examples include flood barriers, storm-resistant buildings, and drought-proof water systems, which are designed to withstand the specific climate risks they face. Moreover, climate-resilient infrastructure enhances the adaptive capacity of communities, enabling them to better cope with and recover from climate impacts. This resilience is crucial not only for immediate disaster response but also for long-term economic stability and growth. When infrastructure can withstand climate shocks, it ensures continuity in essential services such as electricity, water supply, and transportation, which are vital for both daily life and emergency responses.

For example, resilient power grids can prevent widespread outages during heatwaves or storms, maintaining critical operations in hospitals, schools, and businesses. Another significant benefit of climate-resilient infrastructure is the reduction in economic losses. Extreme weather events can cause extensive damage to infrastructure, resulting in substantial repair costs and economic disruptions. By investing in resilience, governments and businesses can avoid these costs and minimize downtime. For instance, coastal cities that invest in seawalls and flood defences can protect their real estate, commercial districts, and industrial zones from storm surges and rising sea levels. This not only saves money on repairs but also preserves the economic vitality of these areas. Furthermore, climate-resilient infrastructure contributes to social equity by protecting vulnerable populations.

Low-income communities often reside in areas that are more susceptible to climate impacts and have less capacity to recover from disasters. Investing in resilient infrastructure in these areas can provide much-needed protection and reduce the disparity in climate impacts. For example, building flood-resistant housing and improving drainage systems in low-lying urban neighbourhoods can protect residents from frequent flooding, improving their quality of life and economic prospects. In addition to physical infrastructure, climate resilience also involves strengthening institutional frameworks and community preparedness. This includes implementing policies that mandate resilience in new developments, retrofitting existing infrastructure, and promoting the use of climate-resilient materials and technologies.

For instance, updating building codes to require higher standards of wind resistance in hurricane-prone areas can significantly reduce damage and enhance safety. Governments can also incentivize the adoption of resilient practices through grants, subsidies, and technical assistance programmes. The environmental benefits of climate-resilient infrastructure are also noteworthy. Nature-based solutions, such as restoring wetlands and planting urban forests, can enhance resilience while providing ecological benefits like biodiversity conservation and carbon sequestration. These green infrastructure projects can absorb floodwaters, reduce urban heat islands, and improve air and water quality, contributing to a healthier environment and more resilient communities. The long-term benefits of investing in climate-resilient infrastructure extend to economic growth and development. Resilient infrastructure attracts investment by providing a stable and secure environment for businesses.

Companies are more likely to invest in regions where infrastructure can withstand climate impacts, ensuring the continuity of operations and reducing the risk of asset loss. This can lead to job creation, increased economic activity, and improved livelihoods for local populations. In conclusion, climate-resilient infrastructure is essential for protecting lives and property, reducing economic losses, and enhancing the adaptive capacity of communities. It plays a critical role in ensuring the long-term sustainability and safety of societies as they face the growing threats posed by climate change. By investing in resilient infrastructure, governments and businesses can safeguard essential services, promote social equity, and support sustainable economic growth. The integration of resilient practices into infrastructure planning and development is not just a necessity but a strategic imperative for building a more secure and prosperous future in the face of climate change.

Investment Strategies and Financial Mechanisms

To effectively build and maintain climate-resilient infrastructure, significant financial investments are required. Various investment strategies and financial mechanisms can be leveraged to mobilize the necessary resources.

Public Investment and Budget Allocation

Governments play a central role in funding climate-resilient infrastructure projects through public investment and budget allocation. By prioritizing resilience in their fiscal planning, governments can ensure that sufficient resources are dedicated to upgrading existing infrastructure and developing new projects that can withstand climate impacts. This might include constructing flood barriers, retrofitting buildings to endure extreme weather, and enhancing drainage systems to manage heavy rainfall. Public investment in climate resilience is crucial for several reasons. First, it addresses market failures where private sector investments might be insufficient due to the long-term nature and high upfront costs of resilience projects.

Governments can use budget allocations to fund critical infrastructure that provides widespread public benefits, such as flood defences, seawalls, and climate-resilient transportation networks. These investments not only protect communities but also reduce the long-term economic costs associated with climate-related damages and disruptions. Moreover, public funding can be directed towards research and development (R&D) to advance resilience technologies and practices. By supporting innovation in construction materials, early warning systems, and sustainable design practices, governments can foster the development of new solutions that enhance infrastructure resilience. Additionally, targeted subsidies and grants can incentivize private sector participation in resilience-building activities, encouraging businesses to adopt climate-smart practices and technologies.

Public-Private Partnerships (PPPs)

Public-Private Partnerships (PPPs) are an effective way to leverage private sector resources and expertise in financing and implementing climate-resilient infrastructure projects. These partnerships can pool public and private funds, share risks, and combine technical capabilities to achieve more efficient and innovative solutions. In a typical PPP arrangement, the public sector provides regulatory support, land, or initial funding, while the private sector contributes capital, technical expertise, and operational management. This collaboration can result in high-quality infrastructure that meets resilience standards and delivers public benefits. For example, a PPP might involve a private company designing and constructing a resilient transportation system, with the government ensuring that the project aligns with national resilience goals and providing necessary regulatory approvals.

PPPs offer several advantages for climate-resilient infrastructure projects. They can attract private investment by reducing the financial burden on public budgets, accelerate project timelines through efficient project management, and foster innovation by integrating cutting-edge technologies and practices. Additionally, PPPs can enhance accountability and performance through clear contractual agreements that specify the responsibilities and expectations of each partner. Successful PPPs require a conducive regulatory environment, transparent procurement processes, and robust governance frameworks. Governments must establish clear guidelines for PPP projects, ensure competitive bidding processes, and monitor project implementation to safeguard public interests. By creating an enabling environment for PPPs, governments can unlock significant private sector contributions to climate resilience.

Climate Bonds

Climate bonds are financial instruments designed to raise capital for projects that deliver environmental benefits, including climate resilience. By issuing climate bonds, governments, development banks, and private entities can attract investment for infrastructure projects that reduce vulnerability to climate impacts. Investors are increasingly interested in green finance, making climate bonds a viable option for funding large-scale resilience projects. Climate bonds function similarly to traditional bonds but are specifically earmarked for environmental projects. The proceeds from these bonds are used to finance projects such as flood defences, renewable energy installations, and sustainable water management systems. The appeal of climate bonds lies in their dual benefit: they provide financial returns to investors while contributing to environmental sustainability and climate resilience.

The issuance of climate bonds can be supported by frameworks such as the Green Bond Principles (GBP) or the Climate Bonds Standard, which provide guidelines for transparency, reporting, and project selection. These frameworks ensure that the funds are used for genuine environmental benefits and enhance investor confidence. Governments and financial institutions can also offer incentives, such as tax exemptions or interest rate subsidies, to attract more investors to climate bonds. Climate bonds are financial instruments designed to raise capital for projects that deliver environmental benefits, including climate resilience. By issuing climate bonds, governments, development banks, and private entities can attract investment for infrastructure projects that reduce vulnerability to climate impacts.

The growing interest in green finance among investors makes climate bonds a viable option for funding large-scale resilience projects. Climate bonds operate similarly to traditional bonds but are specifically earmarked for environmentally beneficial projects. The proceeds from these bonds are used to finance initiatives such as flood defences, renewable energy installations, and sustainable water management systems. These projects are essential for building infrastructure that can withstand the impacts of climate change and ensure long-term sustainability. The attractiveness of climate bonds lies in their dual benefit: they provide financial returns to investors while contributing to environmental sustainability and climate resilience. This dual benefit appeals to a broad range of investors, including institutional investors, green investment funds, and socially responsible investors who are looking to balance financial returns with positive environmental impact.

To ensure transparency and accountability, the issuance of climate bonds is often guided by established frameworks such as the Green Bond Principles (GBP) or the Climate Bonds Standard. These frameworks provide guidelines for project selection, transparency, and reporting, ensuring that the funds rose is genuinely used for projects that deliver environmental benefits. Compliance with these standards enhances investor confidence and increases the attractiveness of climate bonds. Governments and financial institutions can further support the issuance of climate bonds by offering incentives such as tax exemptions or interest rate subsidies. These incentives can lower the cost of issuing bonds, making them more appealing to potential issuers. Additionally, public sector entities can collaborate with international organizations and development banks to co-issue climate bonds, leveraging their credibility and networks to attract more investors.

Insurance and Risk Transfer Mechanisms

Innovative insurance products and risk transfer mechanisms are critical tools for providing financial protection against climate risks. These mechanisms help manage the financial impact of climate-related disasters, ensuring that funds are available for immediate response and recovery efforts.

Parametric Insurance

Parametric insurance is an innovative insurance product that pays out based on predefined triggers, such as specific weather events. Unlike traditional insurance, which requires an assessment of actual damages before payouts can be made, parametric insurance provides quick financial relief once the predefined conditions are met. This rapid payout is crucial for enabling timely repairs and recovery efforts, minimizing the economic disruption caused by climate events. For example, a parametric insurance policy might provide coverage for floods by specifying a payout if rainfall exceeds a certain threshold within a specific period. This setup ensures that funds are available as soon as the trigger conditions are met, without the need for lengthy claims processing. Parametric insurance is particularly useful for covering high-frequency, low-severity events that can cause significant cumulative damage over time.

Catastrophe Bonds (Cat Bonds)

Catastrophe bonds, or cat bonds, are another innovative risk transfer mechanism that allows governments to transfer the risk of severe climate events to investors. These bonds provide a financial safety net for governments, enabling them to fund immediate response and recovery efforts without straining public budgets. Cat bonds are structured so that investors receive periodic interest payments, and the principal is returned at the bond's maturity, provided that a predefined catastrophe event does not occur. If the specified event (such as an earthquake, hurricane, or flood) does occur, the principal is used to cover the financial losses associated with the event, and investors may lose their principal investment.

This structure provides governments with immediate access to funds for disaster response and recovery. The issuance of cat bonds requires careful risk assessment and modelling to determine the likelihood of the specified events and set appropriate trigger conditions. Collaborating with experienced financial institutions and risk modelling experts is essential for designing effective cat bonds. Additionally, governments can enhance the appeal of cat bonds to investors by offering guarantees or co-issuing bonds with international organizations that have strong credit ratings.

Both parametric insurance and cat bonds can be part of a broader risk management strategy that includes traditional insurance, emergency response funds, and contingency budgeting. By diversifying their risk management approaches, governments can ensure they have multiple layers of financial protection against climate risks.

Conclusion

Climate bonds and innovative insurance products are essential financial mechanisms for mobilizing the resources needed to build climate-resilient infrastructure. Climate bonds attract investment for projects that deliver significant environmental benefits, while insurance products and risk transfer mechanisms provide financial protection against the impacts of climate-related disasters. Together, these tools enable governments, development banks, and private entities to fund and implement resilience projects that safeguard communities and promote sustainable development. By leveraging these financial mechanisms, policymakers can enhance the adaptive capacity of infrastructure, reduce vulnerability to climate impacts, and ensure a more resilient and sustainable future. Strategic investments in climate resilience not only protect against immediate threats but also contribute to long-term economic stability and growth, making them a critical component of comprehensive climate adaptation strategies.

Innovation in Building Climate Resilience

Innovation plays a crucial role in enhancing the climate resilience of infrastructure. Advances in technology, materials, and design practices can significantly improve the ability of infrastructure to withstand and adapt to climate impacts. These innovations not only help protect physical structures but also enhance the overall sustainability and functionality of urban and rural environments.

Advanced Materials and Construction Techniques

The development of new materials and construction techniques is vital for building resilient infrastructure. Traditional building materials and methods often fall short when faced with the increasing severity of climate events. Innovations in this area can provide enhanced durability and longevity, thereby reducing the likelihood of damage and the need for costly repairs.

- **High-Performance Concrete and Corrosion Resistant Materials:** High-performance concrete is an example of an advanced material that offers superior strength and durability compared to conventional concrete.

It can withstand extreme weather conditions, such as intense heat and heavy rainfall, making it ideal for use in climate-resilient infrastructure projects. Similarly, corrosion-resistant materials, such as certain alloys and coated steel, can protect infrastructure from the damaging effects of saltwater and humidity, which are common in coastal areas prone to storms and sea-level rise.

- **Modular Construction Techniques:** Modular construction is an innovative approach that involves assembling buildings or infrastructure in sections, or modules, which are then transported to the construction site for final assembly. This technique can significantly speed up the building process, reduce waste, and ensure high standards of quality and resilience. Modular construction allows for better control over the quality of materials and construction processes, ensuring that structures meet rigorous resilience standards.

Smart Infrastructure and IoT

The integration of smart technologies and the Internet of Things (IoT) into infrastructure can enhance climate resilience by enabling real-time monitoring and adaptive management. These technologies provide valuable data and insights that can inform proactive maintenance and emergency response strategies.

- **Sensors and IoT Devices:** Sensors and IoT devices can be embedded in infrastructure to monitor environmental conditions, structural integrity, and usage patterns. For example, sensors in bridges and buildings can detect stress and strain, alerting authorities to potential issues before they become critical. Similarly, weather stations and water level sensors can provide real-time data on rainfall, river levels, and other environmental factors, enabling early warning systems for floods and storms.
- **Smart Water Management Systems:** Smart water management systems use IoT technology to optimize the

distribution and use of water resources. These systems can monitor water consumption, detect leaks, and manage irrigation more efficiently, helping to conserve water during droughts and prevent flooding during heavy rainfall. By integrating real-time data with automated controls, smart water management systems can respond quickly to changing conditions, enhancing resilience to both droughts and floods.

Nature-based Solutions

Incorporating nature-based solutions into infrastructure projects can provide multiple benefits for climate resilience. These solutions leverage natural processes and ecosystems to mitigate climate impacts while also enhancing biodiversity and providing ecosystem services.

- **Urban Green Spaces and Wetlands:** Urban green spaces, such as parks and green roofs, can absorb rainfall, reduce runoff, and mitigate the urban heat island effect. Wetlands, both natural and constructed, can act as buffers against floods by absorbing and storing excess water. These green infrastructure elements not only enhance resilience to climate impacts but also improve air and water quality, support biodiversity, and provide recreational spaces for communities.
- **Mangrove Restoration and Reforestation:** Mangrove forests and other coastal ecosystems play a critical role in protecting shorelines from erosion and storm surges. Restoring mangroves and other coastal habitats can provide a natural defence against rising sea levels and extreme weather events. Similarly, reforestation projects can stabilize soils, enhance water retention, and create carbon sinks, contributing to both climate mitigation and adaptation efforts.

Innovative Design Practices

Design practices that incorporate resilience principles can transform how infrastructure is planned and constructed, ensuring that it is capable of withstanding future climate conditions.

- **Resilient Urban Planning:** Urban planning that prioritizes climate resilience can include zoning regulations that prevent development in high-risk areas, such as floodplains and coastal zones. It can also promote the use of permeable surfaces, green roofs, and other green infrastructure to manage storm water and reduce flood risks. Integrating resilience into urban planning helps create cities that are better prepared for climate impacts.
- **Flexible and Adaptive Designs:** Designing infrastructure with flexibility and adaptability in mind can enhance its resilience to unforeseen climate events. For example, buildings can be designed with adjustable foundations that can be elevated in response to rising sea levels. Transport infrastructure, such as roads and railways, can be designed to withstand temperature fluctuations and heavy rainfall, ensuring that they remain functional during extreme weather events.

Conclusion

Innovation in building climate resilience is essential for protecting infrastructure and ensuring the sustainability of communities in the face of climate change. Advanced materials and construction techniques, smart infrastructure and IoT, nature-based solutions, and innovative design practices all contribute to creating resilient systems capable of withstanding and adapting to climate impacts. By investing in these innovations, governments, businesses, and communities can enhance their adaptive capacity, reduce vulnerabilities, and promote long-term economic and environmental sustainability. These efforts are critical for building a more resilient and climate-adaptive future, safeguarding lives, property, and essential services against the increasing threats posed by climate change.

Smart Infrastructure and IoT

The integration of smart technologies and the Internet of Things (IoT) into infrastructure can significantly enhance climate resilience by enabling real-time monitoring and adaptive management. These advanced systems collect and analyze data on environmental conditions, infrastructure performance, and usage patterns, providing valuable insights that inform proactive maintenance and emergency response strategies.

Sensors and IoT Devices

Sensors and IoT devices play a critical role in monitoring infrastructure and environmental conditions. By embedding sensors in bridges, buildings, and roads, authorities can continuously track the structural integrity and detect early signs of wear and damage. For example, sensors in bridges can monitor stress and strain, providing alerts if the structure is at risk of failure due to extreme weather or aging materials. This early warning system allows for timely maintenance and repairs, preventing catastrophic failures and ensuring public safety.

Smart Grids

Smart grids enhance the reliability and efficiency of energy supply during extreme weather events. These grids use IoT technology to monitor and manage the distribution of electricity, quickly identifying and responding to outages or imbalances. During heat waves or storms, smart grids can reroute power to ensure continuous supply to critical facilities such as hospitals and emergency services. They can also integrate renewable energy sources, such as solar and wind, more effectively, providing a flexible and resilient energy network that can adapt to changing conditions. By improving the management and distribution of energy, smart grids contribute to the overall resilience of urban infrastructure.

Nature-Based Solutions

Incorporating nature-based solutions into infrastructure projects offers multiple benefits for climate resilience. These solutions leverage natural processes and ecosystems to mitigate climate impacts while also enhancing biodiversity and providing ecosystem services.

Green Infrastructure

Green infrastructure, such as urban green spaces, wetlands, and mangroves, acts as a natural buffer against floods and storms. Urban green spaces, including parks and green roofs, absorb rainfall, reducing runoff and mitigating the urban heat island effect. These areas not only enhance resilience but also improve air quality and provide recreational opportunities for residents. Wetlands, both natural and constructed, can absorb floodwaters and filter pollutants, protecting water quality and reducing the risk of flooding in adjacent areas.

Coastal Mangroves

Restoring and conserving coastal mangroves is a particularly effective nature-based solution for protecting shorelines from erosion and storm surges. Mangroves have dense root systems that stabilize the soil and dissipate wave energy, reducing the impact of storms on coastal communities.

This natural protection reduces the need for expensive artificial barriers, such as seawalls, which can be costly to build and maintain. Additionally, mangroves provide critical habitat for marine life, support fisheries, and sequester carbon, contributing to both climate mitigation and adaptation efforts.

Reforestation and Agro-forestry

Reforestation and agro-forestry practices enhance climate resilience by stabilizing soils, increasing water retention, and creating carbon sinks. Planting trees in degraded areas can prevent landslides and reduce runoff, protecting communities and infrastructure from floods. Agro-forestry, which integrates trees and shrubs into agricultural landscapes, enhances biodiversity, improves soil health, and provides shade and shelter for crops and livestock. These practices not only improve agricultural productivity but also increase the resilience of rural communities to climate impacts.

Urban Forestry

Urban forestry involves planting and maintaining trees in urban areas to enhance resilience and improve quality of life. Trees in cities provide shade; reduce temperatures, and lower energy consumption by cooling buildings. They also absorb rainwater, reducing runoff and the risk of flooding. Urban forestry projects can be integrated into city planning to create green corridors, parks, and street trees that contribute to the overall resilience of urban areas. These projects provide aesthetic and recreational benefits, making cities more livable and sustainable.

Conclusion

The integration of smart technologies and IoT into infrastructure, along with the implementation of nature-based solutions, is essential for building climate resilience. Advanced materials, sensors, and smart systems enable real-time monitoring and adaptive management, ensuring that infrastructure can withstand and adapt to changing climate conditions. Nature-based solutions leverage the protective functions of ecosystems to mitigate climate impacts, providing cost-effective and sustainable alternatives to traditional engineering approaches. By investing in smart infrastructure and nature-based solutions, governments, businesses, and communities can enhance their adaptive capacity, reduce vulnerabilities, and promote long-term environmental and economic sustainability.

These innovations not only protect physical structures but also improve the overall functionality and quality of urban and rural environments. Through strategic planning and investment, it is possible to create resilient infrastructure that safeguards lives, property, and essential services against the increasing threats posed by climate change.

Case Studies on Successful Investments

The Netherlands: Delta Works

The Netherlands' Delta Works is a prime example of successful investment in climate-resilient infrastructure. Faced with the constant threat of flooding, the Dutch Government embarked on an ambitious project to safeguard the country from the sea. The Delta Works, a comprehensive system of dams, sluices, locks, dikes, and storm surge barriers, is designed to protect the low-lying regions of the Netherlands from the devastating impacts of storm surges and rising sea levels.

Comprehensive Protection System

The Delta Works project was initiated in response to the catastrophic North Sea Flood of 1953, which resulted in significant loss of life and property. Recognizing the need for robust flood defences, the Dutch Government invested heavily in engineering solutions to enhance the country's resilience to future floods. The system comprises several key components:

- **Dams and Dikes:** These structures form the backbone of the Delta Works, providing physical barriers against the sea. The dams and dikes are built to withstand severe storm surges, ensuring that coastal and inland areas remain protected.
- **Storm Surge Barriers:** Notable among these is the Oosterscheldekering, one of the largest storm surge barriers in the world. This barrier can be closed during extreme weather events to prevent seawater from inundating the land. Its innovative design allows it to remain open most of the time, preserving the natural tidal flow and the local ecosystem.
- **Sluices and Locks:** These control the flow of water and maintain safe water levels within the protected areas. They also facilitate navigation, ensuring that commercial and recreational vessels can move safely through the waterways.

Economic and Social Benefits

The Delta Works project has had far-reaching economic and social benefits. By providing a high level of flood protection, it has created secure conditions for agriculture, industry, and tourism, all of which are vital to the Dutch economy.

- **Agriculture:** The Netherlands is renowned for its agricultural productivity, particularly in areas that lie below sea level. The Delta Works ensures that these fertile lands are protected from flooding, allowing farmers to cultivate crops and raise livestock without the constant threat of inundation. This security has enabled the agricultural sector to thrive, contributing significantly to the country's GDP and food security.
- **Industry:** Many of the Netherlands' industrial activities are located in coastal regions and along major rivers. The Delta Works provides these industries with the confidence to invest and expand, knowing that their infrastructure is protected from the sea. This has bolstered economic growth and job creation in sectors such as manufacturing, logistics, and energy.
- **Tourism:** The Netherlands attracts millions of tourists each year, drawn by its picturesque landscapes, historic cities, and cultural heritage. The Delta Works itself has become a tourist attraction, showcasing Dutch ingenuity and resilience. Secure coastal areas and scenic waterways enhance the visitor experience, supporting the tourism industry and local economies.

Adaptive Management and Continuous Improvement

The success of the Delta Works is attributed not only to its robust engineering but also to its adaptive management practices. The Dutch Government continually monitors and maintains the infrastructure, incorporating the latest climate data and projections to ensure its effectiveness.

- **Continuous Maintenance:** Regular inspections and maintenance activities are conducted to keep the structures in optimal condition. This includes reinforcing dikes, updating mechanical systems in sluices and locks, and repairing any damage caused by wear and tear or extreme weather events.
- **Incorporating Climate Data:** The Delta Works project is designed to be adaptable to changing climate conditions. Engineers and planners use the latest climate models and projections to assess potential future risks, such as sea-level rise and increased storm intensity. This proactive approach allows for timely upgrades and modifications to the infrastructure, ensuring that it remains effective in protecting the Netherlands from evolving climate threats.
- **Innovation and Research:** The Netherlands continues to invest in research and innovation to enhance its flood protection strategies. This includes exploring new materials, construction techniques, and technologies that can improve the resilience and efficiency of the Delta Works. Collaboration with international experts and institutions ensures that the project benefits from global best practices and cutting-edge scientific knowledge.

Conclusion

The Delta Works stands as a testament to the Netherlands' commitment to climate resilience and adaptive management. By investing in comprehensive and innovative flood protection infrastructure, the Dutch government has safeguarded lives and property, supported economic development, and enhanced the country's ability to cope with the impacts of climate change. The project's success highlights the importance of robust engineering, continuous maintenance, and adaptive management in building climate-resilient infrastructure. The Delta Works not only protects the Netherlands today but also provides a model for other countries facing similar challenges, demonstrating how strategic investment in resilience can secure a sustainable and prosperous future.

New York City: East Side Coastal Resiliency Project

In response to the devastation caused by Hurricane Sandy in 2012, New York City launched the East Side Coastal Resiliency Project (ESCR), a comprehensive flood protection initiative aimed at protecting vulnerable neighbourhoods along the East River. This project embodies an innovative approach to urban resilience by combining traditional hard infrastructure with nature-based solutions to enhance the city's ability to withstand and recover from severe weather events.

Comprehensive Flood Protection

The East Side Coastal Resiliency Project is designed to address the significant flood risks faced by neighbourhoods on Manhattan's east side. These areas are particularly vulnerable to storm surges and rising sea levels, which were starkly highlighted by the impacts of Hurricane Sandy. The project encompasses several key components to provide robust flood protection:

- **Floodwalls and Levees:** The construction of floodwalls and levees forms the backbone of the ESCR. These structures are designed to prevent storm surges from inundating the neighbourhoods, safeguarding homes, businesses, and critical infrastructure. The floodwalls are engineered to withstand extreme weather events, ensuring long-term protection against future storms.
- **Green Spaces:** Integrating green spaces into the flood protection strategy are a central element of the ESCR. Parks, gardens, and other green areas are designed to absorb storm water, reduce runoff, and provide recreational opportunities for residents. These green spaces act as natural buffers, complementing the hard infrastructure and enhancing the overall resilience of the urban environment.

Integration of Hard Infrastructure and Nature-Based Solutions

One of the defining features of the ESCR is its integration of hard infrastructure with nature-based solutions. This holistic approach not only strengthens flood defences but also offers additional environmental and social benefits:

- **Storm Surge Barriers:** Strategically placed storm surge barriers are deployed to protect against high water levels. These barriers can be activated during extreme weather events to prevent flooding while remaining unobtrusive during normal conditions, preserving the aesthetic and functional qualities of the waterfront.
- **Bioswales and Rain Gardens:** The project incorporates bioswales and rain gardens, which are designed to manage stormwater runoff. These features use vegetation and soil to filter and absorb rainwater, reducing the burden on the city's drainage systems and minimizing the risk of flooding. They also enhance urban biodiversity and improve air quality.
- **Waterfront Parks:** New waterfront parks are being developed as part of the ESCR, providing open spaces for recreation and community activities. These parks are designed with resilient landscaping that can endure flooding and recover quickly, ensuring that they remain accessible and functional even after extreme weather events.

Community and Environmental Benefits

The ESCR goes beyond flood protection by delivering substantial recreational and environmental benefits to the community. By creating green spaces and enhancing public amenities, the project aims to improve the quality of life for residents and foster a sense of community ownership and stewardship.

- **Recreational Opportunities:** The new parks and green spaces provide residents with places to relax, exercise, and connect with nature. These areas offer a range of recreational facilities, including playgrounds, sports fields, and walking paths, promoting physical activity and well-being.

- **Environmental Improvements:** The integration of green infrastructure enhances the local environment by promoting biodiversity, improving air and water quality, and creating habitats for urban wildlife. The project also contributes to the city's broader sustainability goals by incorporating climate adaptation and mitigation measures.
- **Community Engagement:** Community engagement has been a critical component of the ESCR. Throughout the planning and implementation phases, city officials have worked closely with local residents, businesses, and stakeholders to ensure that the project meets the needs and expectations of the community. Public workshops, meetings, and consultations have provided opportunities for feedback and collaboration, fostering a sense of ownership and accountability.

Adaptive Management and Future Resilience

The success of the ESCR relies on adaptive management practices that ensure the project remains effective in the face of evolving climate conditions. This includes continuous monitoring, maintenance, and updates based on the latest climate data and projections.

- **Continuous Monitoring:** Sensors and monitoring systems are installed throughout the project area to track environmental conditions, infrastructure performance, and potential vulnerabilities. This data informs proactive maintenance and allows for timely interventions to address any emerging issues.
- **Regular Maintenance:** Maintaining the flood protection infrastructure and green spaces is crucial for long-term resilience. The city has established protocols for regular inspections, repairs, and upgrades to ensure that all components remain in optimal condition and ready to respond to future climate events.

- **Incorporating New Data:** The ESCR is designed to be flexible and adaptable, incorporating new scientific data and climate projections to update and improve resilience measures. This adaptive approach ensures that the project remains relevant and effective as climate conditions change over time.

Conclusion

The East Side Coastal Resiliency Project exemplifies New York City's commitment to enhancing urban resilience through innovative and integrated approaches. By combining hard infrastructure with nature-based solutions, the ESCR provides comprehensive flood protection while delivering significant recreational and environmental benefits to the community. The project's success demonstrates the importance of adaptive management, continuous community engagement, and strategic investment in building resilient cities that can withstand and thrive in the face of climate change. Through such forward-thinking initiatives, New York City is setting a benchmark for urban resilience that can inspire and guide other cities around the world.

Singapore: Smart Water Management

Singapore has made substantial investments in smart water management systems to enhance its resilience to both floods and droughts. The city-state's innovative approach combines green infrastructure with advanced water management technologies, demonstrating a holistic strategy to manage storm water, improve water quality, and enhance the urban environment. The Active, Beautiful, Clean (ABC) Waters Programme exemplifies this integrated approach, turning water management into a multifunctional, sustainable system.

Active, Beautiful, Clean (ABC) Waters Programme

The ABC Waters Programme is a cornerstone of Singapore's water management strategy. Launched by the Public Utilities Board

(PUB), the programme aims to transform Singapore's waterways and water bodies into vibrant, clean, and accessible spaces that contribute to the city's resilience and liveability. By integrating green infrastructure with advanced technologies, the programme addresses both flood management and water quality improvement.

Green Infrastructure

Green infrastructure is a key element of the ABC Waters Programme, providing natural solutions to storm water management. Key features include:

- **Rain Gardens:** These landscaped areas are designed to absorb and filter rainwater. By using a mix of soil, sand, and vegetation, rain gardens slow down runoff, enhance infiltration, and improve water quality by trapping pollutants before they reach the waterways.
- **Detention Ponds:** These ponds temporarily store stormwater during heavy rainfall, reducing peak flow rates and mitigating flood risks. Detention ponds are often designed as attractive water features that enhance urban aesthetics and provide recreational spaces for residents.
- **Bio-retention Swales:** Swales are shallow, vegetated channels that convey, filter, and infiltrate storm water. Bio-retention swales are particularly effective in urban areas, where they can be integrated into streetscapes, parks, and parking lots, reducing surface runoff and improving water quality.

Advanced Water Management Technologies

The integration of advanced technologies enhances the efficiency and effectiveness of Singapore's water management systems. Smart sensors and real-time monitoring systems play a crucial role in this regard:

- **Smart Sensors:** Deployed throughout the city's water infrastructure, these sensors collect data on rainfall, water levels, and flow rates. This real-time data is critical for monitoring the performance of water management systems and identifying potential issues before they escalate into major problems.

- **Real-Time Monitoring Systems:** These systems aggregate data from various sensors and provide a comprehensive overview of the water management network. By enabling continuous monitoring, real-time systems allow for quick responses to changing conditions, such as sudden heavy rainfall or rising water levels.
- **Automated Controls:** Integrating automated controls with real-time monitoring allows for dynamic management of water resources. For instance, gates and valves in drainage systems can be automatically adjusted to optimize the flow of storm water, preventing overflows and reducing flood risks.

Enhancing Flood Resilience

Singapore's smart water management systems significantly enhance its resilience to floods. By capturing and treating storm water close to its source, green infrastructure elements like rain gardens and Bio-retention swales reduce the volume and speed of runoff, decreasing the likelihood of urban flooding. Detention ponds and other storage features further mitigate flood risks by holding excess water during peak rainfall events. Moreover, the integration of smart technologies ensures that the city's drainage systems are managed efficiently.

Real-time data allows for predictive modelling and early warning systems, enabling authorities to take proactive measures in anticipation of extreme weather events. This approach not only protects property and infrastructure but also safeguards lives.

Addressing Droughts and Water Quality

Singapore's water management strategy also addresses the challenges posed by droughts and ensures high water quality. The city-state relies heavily on its ability to efficiently manage its water resources, given its limited natural freshwater sources.

- **Water Conservation:** The ABC Waters Programme promotes water conservation through public awareness campaigns and community engagement. Educating residents about the importance of water conservation and the benefits of green infrastructure fosters a culture of sustainability and resilience.
- **Water Reuse:** Singapore has invested in advanced water treatment facilities that enable the reuse of wastewater. Treated wastewater, known as NEWater, is a critical component of Singapore's water supply, providing a reliable source of high-quality water during dry periods.
- **Pollution Control:** Green infrastructure elements not only manage storm water but also improve water quality by filtering out pollutants. This dual functionality is essential for maintaining clean waterways and protecting aquatic ecosystems.

Community and Environmental Benefits

The ABC Waters Programme goes beyond technical solutions, integrating community engagement and environmental stewardship into its framework. By transforming utilitarian water management infrastructure into beautiful, multifunctional spaces, the programme enhances the urban environment and improves the quality of life for residents.

- **Recreational Spaces:** Green infrastructure elements like rain gardens, Bio-retention swales, and detention ponds are designed as attractive features that provide recreational opportunities. Parks and greenways along waterways offer spaces for leisure, exercise, and community gatherings.
- **Biodiversity:** By creating natural habitats within urban areas, the programme supports biodiversity and fosters a healthier ecosystem. Green spaces attract wildlife, enhance

urban greenery, and contribute to the ecological richness of the city.

Conclusion

Singapore's investment in smart water management systems through the ABC Waters Programme demonstrates a forward-thinking approach to urban resilience. By integrating green infrastructure with advanced technologies, the city-state effectively manages storm water, reduces flood risks, and ensures water quality. The combination of innovative solutions and community engagement has transformed Singapore's waterways into active, beautiful, and clean spaces that enhance the city's resilience and liveability. This comprehensive strategy serves as a model for other cities facing similar challenges, highlighting the importance of smart infrastructure and nature-based solutions in building climate resilience.

Investing in climate-resilient infrastructure and innovation is essential for mitigating the impacts of climate change and ensuring the long-term sustainability and safety of communities. By prioritizing resilience in public investment, leveraging private sector resources through PPPs, and utilizing financial mechanisms like climate bonds and insurance, governments can mobilize the necessary resources to build and maintain resilient infrastructure. Innovation in materials, construction techniques, smart technologies, and nature-based solutions further enhances the ability of infrastructure to withstand and adapt to climate impacts. Successful case studies from around the world demonstrate the effectiveness of these investments in protecting communities and fostering sustainable development. Through coordinated efforts and strategic investments, it is possible to build a more resilient and climate-adaptive future.

REFERENCES

Auerbach, A.J. and Gorodnichenko, Y. (2022). "Fiscal Multipliers in Recessions and Expansions: Does It Matter Whether Government Spending is Increasing or Decreasing?" *Journal of Economic Perspectives, 36*(1): 41-64.

Clarida, R.H. (2022). "Monetary Policy, Price Stability, and Inflation: Theoretical Perspectives and Practical Lessons". *Journal of Economic Literature, 60*(1): 34-62.

Coibion, O., Gorodnichenko, Y. and Weber, M. (2022). "How Inflation Expectations Affect Households and Firms: The Role of Central Bank Communication". *Brookings Papers on Economic Activity, 2022*(1): 115-174.

Feldstein, M. (2022). "Inflation, Tax Rules, and the Price of Land and Gold". *Journal of Public Economics*, Vol. 210, pp. 1-294, Editorial Board Article No. 104666.

Gali, J. (2022). *Monetary Policy, Inflation, and the Business Cycle: An Introduction to the New Keynesian Framework*, 3rd edn., Princeton University Press.

Gopinath, G. (2022). "The International Dimension of Inflation: Evidence and Policy Implications". *IMF Staff Papers, 69*(2): 189-218.

Guvenen, F., Kaplan, G. and Song, J. (2022). "How Does Inflation Affect Different Income Groups?" *Journal of Political Economy, 130*(5): 1437-1473.

Hamilton, J.D. (2022). "Why You Should Never Use the Hodrick-Prescott Filter". *Review of Economic Studies, 89*(1): 104-120.

Jordà, Ò. and Taylor, A.M. (2022). "The Pandemic, Inflation, and Debt Sustainability". *Economic Policy, 37*(110): 275-314.

Kiley, M.T. (2022). "Unemployment, Inflation, and Monetary Policy: Evidence from Advanced and Emerging Market Economies". *Journal of International Money and Finance*, Vol. 122, pp. 1-256, Editorial Board Article No. 102656.

Krugman, P. (2022). "Inflation and the Power of Economic Ideas". *Foreign Affairs, 101*(2): 23-37.

Lane, P.R. (2022). "Global Supply Chains and Inflation: The Role of the Pandemic and Policy Responses". *IMF Economic Review, 70*(2): 121-143.

Laubach, T. and Williams, J.C. (2022). "Measuring the Natural Rate of Interest: International Trends and Determinants". *Brookings Papers on Economic Activity, 2022*(3): 57-108.

Leduc, S. and Wilson, D. (2022). "What is Driving the Rise in Inflation? A Primer". *Economic Letter*, Federal Reserve Bank of San Francisco, 2022-27.

Martin, F. (2022). "Inflation, Wage Setting and the Monetary Transmission Mechanism". *Journal of Economic Theory*, Vol. 197, pp. 105-197, Editorial Board Article No. 105299.

Mian, A., Sufi, A. and Verner, E. (2022). "Household Debt and Inflation: The Role of Mortgage Debt in Monetary Policy Transmission". *Journal of Financial Economics, 143*(3): 1041-1061.

Mishkin, F.S. (2022). "Monetary Policy Strategy: Lessons from the Financial Crisis". *Journal of Economic Perspectives, 36*(3): 123-148.

Mussa, M. (2022). "Inflation, Real Interest Rates, and the Real Exchange Rate: An Empirical Investigation". *Journal of Monetary Economics, 127*: 243-271.

Orphanides, A. (2022). "Policy Rules and Economic Stability: Lessons from the Great Inflation". *Journal of Economic Dynamics and Control*, Vol. 135, Editorial Board Article No. 104313.

Reis, R. (2022). "The Analytics of Monetary Policy: A New Keynesian Perspective". *Quarterly Journal of Economics, 137*(4): 1951-1987.

Sims, C.A. (2022). "Fiscal Policy, Inflation, and Debt Sustainability". *Journal of Economic Theory, 198*.

Smets, F. and Wouters, R. (2022). "An Estimated Dynamic Stochastic General Equilibrium Model of the Euro Area". *Journal of Economic Dynamics and Control*, Vol. 133, Editorial Board Article No. 104295.

Stock, J.H. and Watson, M.W. (2022). "Disentangling the Channels of the 2021-2022 Inflation Surge". *Journal of Applied Econometrics, 37*(6): 975-996.

6

Agriculture and Food Security in a Warming World

Economic Implications

Impact of Climate Change on Agriculture

Climate change significantly impacts agriculture, posing substantial risks to global food security. Rising temperatures, changing precipitation patterns, and the increased frequency and severity of extreme weather events affect crop yields, livestock productivity, and the overall viability of agricultural systems.

Temperature Increases

Higher temperatures can reduce crop yields by accelerating crop maturation, reducing the time for grain filling, and increasing the risk of heat stress. For example, crops such as wheat, rice, and maize are sensitive to temperature spikes during their critical growth stages, leading to lower yields. Additionally, livestock can suffer from heat stress, which impacts their growth, reproduction, and milk production. Higher temperatures also exacerbate the proliferation of pests and diseases that affect crops and livestock. Warmer conditions can expand the range and lifespan of pests like aphids and locusts, as well as pathogens that cause crop diseases such as rusts and blights. This increases the need for pest control measures, raising production costs and potentially leading to the overuse of pesticides, which can harm the environment.

Changing Precipitation Patterns

Altered rainfall patterns can lead to both droughts and floods, presenting significant challenges for agriculture. Droughts can cause soil degradation, reduce water availability for irrigation, and lead to crop failures. Prolonged dry spells can deplete soil moisture and reduce groundwater levels, making it difficult for farmers to sustain crop growth. In rainfed agriculture, which relies directly on precipitation, inconsistent rainfall can disrupt planting and harvesting schedules, reducing yields and productivity. Conversely, excessive rainfall and flooding can damage crops, erode soils, and disrupt agricultural infrastructure. Flooded fields can lead to root rot and other water logging-related issues, severely impacting crop health and reducing yields.

Infrastructure such as roads and irrigation canals can be damaged by floods, hindering access to markets and inputs. The recovery from such events can be lengthy and costly, further straining the resources of affected farmers.

Extreme Weather Event

The increased frequency of extreme weather events, such as hurricanes, cyclones, and hailstorms, can devastate agricultural regions. These events can destroy crops, damage infrastructure, and disrupt supply chains, leading to immediate food shortages and long-term economic losses. For example, hurricanes can flatten fields of crops and cause widespread destruction to farm buildings, machinery, and storage facilities. The recovery process can be prolonged, and the economic impact can ripple through communities dependent on agriculture for their livelihoods. Additionally, extreme weather events can exacerbate the vulnerability of agricultural systems already under stress from other climate impacts. The combination of heat waves and droughts, for instance, can be particularly damaging, as crops and livestock are simultaneously deprived of water and subjected to high temperatures. These compounded stresses can lead to significant losses in agricultural productivity and food security.

Sea-Level Rise

Sea-level rise poses a particular threat to coastal agricultural areas. Saltwater intrusion into freshwater systems can render irrigation water unusable, and the inundation of agricultural land can lead to the loss of arable land. Coastal regions that rely on rice paddies or other low-lying crop systems are especially vulnerable. The salinization of soil can reduce its fertility and hinder crop growth, necessitating costly mitigation measures such as soil amendments and desalination.

Carbon Dioxide Levels

While increased levels of atmospheric carbon dioxide can enhance photosynthesis and potentially boost crop yields, this benefit is often offset by the negative effects of other climate change factors such as temperature increases and water stress. Moreover, elevated CO_2 levels can reduce the nutritional quality of some staple crops. For instance, studies have shown that higher CO_2 concentrations can lower the protein and mineral content of wheat and rice, which are critical food sources for millions of people.

Conclusion

The impacts of climate change on agriculture are multifaceted and complex, affecting all aspects of agricultural production from crop yields to livestock productivity. Temperature increases, changing precipitation patterns, extreme weather events, sea-level rise, and elevated carbon dioxide levels all pose significant challenges to global food security. Addressing these challenges requires comprehensive adaptation strategies that integrate climate-smart agricultural practices, technological innovations, and sustainable resource management. By understanding and mitigating the impacts of climate change on agriculture, we can work towards ensuring a resilient and secure food system for the future.

Economic Challenges for Food Security

The economic challenges posed by climate change to food security are multifaceted, affecting production costs, market stability, and access to food. These challenges highlight the complex interplay between environmental changes and economic systems, emphasizing the need for comprehensive strategies to ensure food security in a warming world.

Increased Production Costs

Climate change can significantly increase the costs of agricultural production. Farmers must adapt to changing conditions by investing in irrigation systems, pest control measures, and resilient crop varieties, among other adaptations. These additional costs can be prohibitive, particularly for small-scale and resource-poor farmers, reducing their ability to sustain livelihoods and contribute to food security.

- **Irrigation Systems:** In regions where rainfall patterns have become unpredictable, the installation of efficient irrigation systems is essential to ensure a stable water supply for crops. However, the initial investment in irrigation infrastructure, such as drip irrigation or sprinkler systems, can be substantial. Additionally, ongoing maintenance and operational costs, including energy for pumping water, can further strain farmers' budgets. For small-scale farmers, who often operate on thin profit margins, these costs can be particularly burdensome, potentially leading to reduced production or even abandonment of farming.
- **Pest Control Measures:** Rising temperatures and changing precipitation patterns can lead to an increase in pest populations and the spread of plant diseases. Farmers may need to invest more in pesticides and other pest control

measures to protect their crops. However, these measures are not only costly but can also lead to increased resistance among pests, necessitating the use of even more potent and expensive chemicals. Moreover, excessive use of pesticides can harm beneficial insects and the environment, leading to long-term sustainability issues.

- **Resilient Crop Varieties:** Developing and adopting resilient crop varieties that can withstand extreme weather conditions, pests, and diseases are critical for maintaining agricultural productivity. However, the research and development required to create these varieties are time-consuming and costly. Once developed, the seeds of these resilient varieties are often more expensive than traditional seeds, adding another layer of cost for farmers. Furthermore, the transition to new crop varieties can involve a learning curve and additional investments in knowledge and training.

Market Volatility

Climate-induced disruptions in agricultural production can lead to significant market volatility. Fluctuations in crop yields and livestock productivity can cause price spikes, making food less affordable for vulnerable populations. Market instability can also discourage investment in agriculture, further threatening food security.

- **Price Spikes:** When climate events such as droughts or floods reduce crop yields, the supply of food products decreases, leading to higher prices. These price spikes can be particularly devastating for low-income households, who spend a large proportion of their income on food. Higher food prices can force these households to reduce their food intake or switch to less nutritious options, exacerbating food insecurity and malnutrition.

- **Investment Discouragement:** Market volatility can create an uncertain environment for agricultural investments. Farmers and investors may be hesitant to invest in new technologies, infrastructure, or expansion if they are unsure about future returns. This hesitation can lead to underinvestment in the agricultural sector, slowing down innovation and productivity improvements that are crucial for adapting to climate change.
- **Supply Chain Disruptions:** Climate change can also disrupt agricultural supply chains, further contributing to market volatility. Extreme weather events can damage transportation infrastructure, delay shipments, and increase logistics costs. These disruptions can lead to temporary shortages of food products in markets, driving up prices and creating panic buying and hoarding behaviours, which further exacerbate price volatility.

Access to Food

Climate change can exacerbate existing inequalities in food access. Vulnerable populations, including low-income households and communities in developing countries, are particularly at risk. As food prices rise and production becomes more unpredictable, these groups may struggle to afford or access sufficient and nutritious food, leading to increased food insecurity and malnutrition.

- **Income Inequality:** Climate change impacts tend to disproportionately affect low-income and marginalized communities, who often lack the resources to adapt. As agricultural productivity declines and food prices rise, these communities face greater challenges in accessing affordable and nutritious food. This can lead to a cycle of poverty and food insecurity, where poor nutrition undermines health and productivity, further reducing income and access to food.
- **Geographic Disparities:** Climate change can exacerbate geographic disparities in food security. Regions that are

more vulnerable to climate impacts, such as Sub-Saharan Africa and South Asia, may experience greater declines in agricultural productivity. These regions are also home to many low-income populations who rely heavily on agriculture for their livelihoods. As local food production becomes more challenging, these communities may become increasingly dependent on imported food, which can be more expensive and less accessible.

- **Nutritional Quality:** Climate change can also affect the nutritional quality of food. Elevated CO_2 level can reduce the concentration of essential nutrients, such as zinc and iron, in staple crops like wheat and rice. This reduction in nutritional quality can have significant health implications, particularly for populations that rely on these staples as their primary food sources. Combined with higher food prices and reduced access, this can lead to widespread malnutrition and related health issues.
- **Government and Institutional Challenges:** Governments and Institutions face significant challenges in addressing the economic impacts of climate change on food security. Effective policy responses require substantial resources, coordination, and long-term commitment, which can be difficult to achieve.
- **Policy and Regulation:** Developing and implementing policies that promote climate resilience in agriculture can be complex and resource-intensive. Governments need to balance immediate food security concerns with long-term sustainability goals. This requires comprehensive planning, stakeholder engagement, and the ability to enforce regulations and support adaptation measures.
- **Funding and Investment:** Securing adequate funding for climate adaptation in agriculture is a major challenge. While international organizations and donors provide some support, the scale of the required investments often exceeds available resources. Innovative financing mechanisms,

such as climate bonds and public-private partnerships, are needed to bridge the funding gap.

- **Research and Development:** Investing in research and development (R&D) is critical for developing new agricultural technologies and practices that can enhance resilience. However, R&D efforts require sustained funding and collaboration between governments, research institutions, and the private sector. Ensuring that the benefits of new technologies are accessible to small-scale and resource-poor farmers is also a key challenge.
- **Extension Services and Education:** Providing farmers with the knowledge and skills needed to adapt to climate change is essential. Effective extension services can disseminate information about climate-smart agricultural practices, resilient crop varieties, and efficient resource management. However, these services often face challenges such as limited funding, insufficient staffing, and difficulties in reaching remote and marginalized communities.

Conclusion

The economic challenges posed by climate change to food security are complex and multifaceted. Increased production costs, market volatility, and access disparities all threaten the stability of food systems and the livelihoods of millions of people. Addressing these challenges requires a comprehensive and coordinated approach that includes substantial investments in climate adaptation, innovative financing mechanisms, effective policy responses, and robust extension services.

Market Volatility

Climate-induced disruptions in agricultural production can lead to significant market volatility. These disruptions, driven by fluctuations in crop yields and livestock productivity, can cause dramatic price spikes, making food less affordable for vulnerable populations and destabilizing markets. Market instability can also discourage investment in agriculture, further threatening food security and economic stability.

Fluctuations in Crop Yields and Livestock Productivity

Climate change affects agricultural production through a variety of mechanisms, including extreme weather events, altered precipitation patterns, and increased temperatures. These changes can result in reduced crop yields and lower livestock productivity. For example, a severe drought can decimate crop fields and reduce the availability of feed for livestock, while extreme heat can lead to livestock mortality and decreased milk production. Such fluctuations create uncertainty in food supply, leading to volatile market conditions.

Price Spikes

When agricultural production is disrupted, the supply of food products decreases, which can lead to significant price increases. These price spikes can be particularly harmful to low-income households that spend a large proportion of their income on food. In developing countries, where food costs can account for a substantial share of household expenses, rising prices can push many families into poverty and hunger. Even in developed countries, sudden increases in food prices can strain household budgets and reduce disposable income for other necessities.

Impact on Vulnerable Populations

Vulnerable populations, including low-income households, rural communities, and small-scale farmers, are disproportionately affected by market volatility. These groups often lack the financial resources and access to credit necessary to cope with rising food prices. As a result, they may be forced to reduce their food intake or switch to less nutritious options, leading to health problems and decreased quality of life. Market volatility can also exacerbate existing inequalities, as wealthier individuals and businesses are better able to absorb price increases and maintain food security.

Investment Discouragement

Market volatility creates an uncertain environment for agricultural investment. Farmers and investors may be hesitant to invest in new technologies, infrastructure, or expansion if they are unsure about future returns. This hesitation can lead to underinvestment in the agricultural sector, slowing down innovation and productivity improvements that are crucial for adapting to climate change. In addition, volatile markets can make it difficult for farmers to plan for the future and manage their operations effectively, further undermining their ability to sustain production and contribute to food security.

Supply Chain Disruptions

Climate-induced disruptions in agricultural production can also affect supply chains, further contributing to market volatility. Extreme weather events can damage transportation infrastructure, delay shipments, and increase logistics costs. These disruptions can lead to temporary shortages of food products in markets, driving up prices and creating panic buying and hoarding behaviours, which further exacerbate price volatility. Supply chain instability can also undermine the reliability of food distribution systems, making it more difficult for consumers to access affordable and nutritious food.

Global Implications

Market volatility in one region can have global repercussions, as the interconnected nature of the global food system means that disruptions in one area can affect food availability and prices worldwide. For example, a severe drought in a major grain-producing region can lead to higher prices for staple crops like wheat and maize, affecting food security in importing countries. This interconnectedness highlights the need for coordinated international efforts to address the impacts of climate change on agriculture and food security.

By taking these steps, it is possible to build resilient agricultural systems that can withstand the impacts of climate change and ensure food security for all.

Adaptation Strategies for Sustainable Agriculture

To mitigate the impacts of climate change on agriculture and ensure food security, comprehensive adaptation strategies must be implemented. These strategies involve enhancing agricultural practices, improving resource management, and strengthening resilience at the community level. Effective adaptation requires a multifaceted approach that addresses the diverse challenges posed by a changing climate.

Climate-Smart Agriculture (CSA)

Climate-Smart Agriculture (CSA) integrates climate change adaptation and mitigation into agricultural systems. This approach aims to increase agricultural productivity, enhance resilience to climate impacts, and reduce greenhouse gas emissions. Key CSA practices include diversification, conservation agriculture, and efficient water management.

- **Diversification:** Diversifying crops and livestock is a crucial strategy for reducing the risk of total loss due to climate events. By cultivating a variety of crops with different climate sensitivities, farmers can ensure that at least some of their crops will survive adverse conditions. For example, drought-tolerant crops can be planted alongside those that require more water, spreading the risk of failure. Additionally, integrating agro-forestry practices, which combine trees and shrubs with crops and livestock, can enhance resilience by providing shade, reducing soil erosion, and improving biodiversity. Agro-forestry systems can also offer additional sources of income, such as fruit, nuts, and timber, contributing to economic stability.
- **Conservation Agriculture:** Conservation agriculture techniques, such as minimum tillage, cover cropping and

crop rotation, improves soil health, reduce erosion, and enhance water retention. Minimum tillage reduces soil disturbance, maintaining soil structure and increasing organic matter. Cover cropping involves planting non-cash crops during off-seasons to protect and enrich the soil. These cover crops can suppress weeds, reduce erosion, and improve soil fertility. Crop rotation involves growing different types of crops in the same area in sequential seasons, which can disrupt pest cycles, reduce soil depletion, and improve overall soil health. These practices collectively increase the resilience of farming systems to climatic variability, ensuring sustainable agricultural production.

- **Efficient Water Management:** Implementing efficient irrigation systems, such as drip irrigation and rainwater harvesting, is vital for optimizing water use and reducing vulnerability to drought. Drip irrigation delivers water directly to the plant roots, minimizing evaporation and ensuring that crops receive adequate water. Rainwater harvesting involves collecting and storing rainwater for agricultural use, providing a reliable water source during dry periods. Additionally, constructing water storage facilities, such as ponds and reservoirs, can enhance water availability and mitigate the impacts of irregular rainfall. Sustainable water management practices ensure that crops receive sufficient water even during dry periods, enhancing resilience to water scarcity.

Improved Resource Management

Sustainable management of natural resources is critical for agricultural adaptation. This includes maintaining soil health, conserving water, and preserving biodiversity. By managing resources sustainably, farmers can create resilient agricultural systems that can withstand climate shocks and stresses.

- **Soil Health:** Maintaining and improving soil fertility through organic farming, composting, and the use of bio-fertilizers can enhance crop resilience and productivity. Organic farming avoids synthetic chemicals, using natural methods to improve soil health and reduce environmental impact. Composting recycles organic waste into valuable soil amendments, enriching the soil with nutrients and improving its structure. Bio-fertilizers, which contain living microorganisms, promote plant growth by increasing nutrient availability and enhancing soil health. Healthy soils are better able to retain water, resist erosion, and support robust plant growth, making them more resilient to climate change.
- **Water Conservation:** Protecting and restoring watersheds, wetlands, and other water bodies helps maintain water availability and quality, which is vital for agriculture. Watersheds collect and store water, releasing it slowly over time to maintain stream flows and groundwater levels. Wetlands act as natural sponges, absorbing excess water during floods and releasing it during dry periods. Restoring and protecting these ecosystems can enhance water security, reduce flood risks, and support agricultural production. Additionally, implementing practices such as contour farming and terracing can reduce runoff and increase water infiltration, further conserving water resources.
- **Biodiversity Conservation:** Preserving genetic diversity in crops and livestock breeds enhances resilience by providing a broader base of traits that can withstand climate stresses. Biodiversity can buffer against environmental changes, pests, and diseases, ensuring stable agricultural production. Conserving traditional crop varieties and livestock breeds that are adapted to local conditions can provide valuable genetic resources for breeding resilient crops and animals. Additionally, integrating natural habitats and wildlife corridors into agricultural landscapes can support beneficial

species, such as pollinators and natural pest predators, further enhancing ecosystem resilience.

Community-Based Adaptation

Engaging local communities in adaptation planning and implementation ensures that strategies are context-specific and culturally appropriate. Community-based adaptation involves participatory planning, capacity building, and the use of local knowledge and resources.

- **Participatory Planning:** Involving farmers and local stakeholders in decision-making processes ensures that adaptation strategies meet their needs and leverage local knowledge. Participatory planning fosters a sense of ownership and commitment, increasing the likelihood of successful implementation. Community meetings, workshops, and collaborative projects can facilitate dialogue and knowledge exchange, empowering communities to develop and implement effective adaptation measures.
- **Capacity Building:** Providing training and resources to farmers helps them adopt new practices and technologies. Extension services, farmer field schools, and demonstration projects can disseminate information about climate-smart agricultural practices, resilient crop varieties, and efficient resource management. Building local capacity enhances the ability of farmers to adapt to climate change, improving their resilience and productivity. Training programmes can also foster innovation and entrepreneurship, encouraging farmers to develop and adopt new solutions.
- **Use of Local Knowledge and Resources:** Leveraging local knowledge and resources can enhance the effectiveness of adaptation strategies. Traditional farming practices, indigenous crop varieties, and local resource management techniques can provide valuable insights and solutions for climate adaptation. Recognizing and integrating this

knowledge into adaptation planning can improve the relevance and sustainability of adaptation measures.

Additionally, promoting local seed banks, community nurseries, and cooperative resource management can support community resilience and self-reliance.

Conclusion

Adaptation strategies for sustainable agriculture are essential for mitigating the impacts of climate change and ensuring food security. By integrating climate-smart agricultural practices, improving resource management, and strengthening community resilience, farmers can build robust agricultural systems that can withstand climate stresses. These strategies not only enhance agricultural productivity but also promote environmental sustainability and social well-being. Effective adaptation requires a holistic and inclusive approach, involving diverse stakeholders and leveraging local knowledge and resources. Through concerted efforts and innovative solutions, it is possible to create resilient agricultural systems that can secure food supplies and support sustainable development in a changing climate.

Improved Resource Management

Sustainable management of natural resources is critical for agricultural adaptation to climate change. Effective resource management strategies help build resilient agricultural systems that can withstand environmental stresses and ensure long-term productivity and food security. Key strategies include maintaining soil health, conserving water, and preserving biodiversity.

Soil Health

Maintaining and improving soil fertility is essential for crop resilience and productivity. Healthy soils support robust plant growth, retain water effectively, and resist erosion, all of which are vital in the face of climate change.

- **Organic Farming:** Organic farming practices avoid synthetic chemicals, relying instead on natural methods to enhance soil health and fertility. Techniques such as crop rotation, cover cropping, and the use of green manures enrich the soil with organic matter and nutrients, improving its structure and fertility. Organic farming also promotes biodiversity, which can help control pests and diseases naturally.
- **Composting:** Composting recycles organic waste into valuable soil amendments. Compost adds nutrients to the soil, improves soil structure, and enhances its ability to retain moisture. This is particularly important in regions prone to drought, as healthy, well-structured soils can store water more efficiently, ensuring that crops receive adequate moisture even during dry periods.
- **Bio-fertilizers:** Bio-fertilizers contain living microorganisms that promote plant growth by increasing nutrient availability in the soil. These microorganisms fix atmospheric nitrogen, solubilise phosphorus, and decompose organic matter, making nutrients more accessible to plants. The use of bio-fertilizers can reduce the need for synthetic fertilizers, which can be costly and environmentally damaging.

Water Conservation

Water conservation is vital for maintaining agricultural productivity, especially in areas affected by changing precipitation patterns and water scarcity. Sustainable water management practices help ensure that crops receive adequate water while minimizing waste and environmental impact.

- **Protecting and Restoring Watersheds:** Watersheds are critical for collecting and storing water, which can be used for irrigation and other agricultural needs. Protecting and restoring watersheds involves conserving forests and vegetation cover, preventing soil erosion, and maintaining natural water flow. Healthy watersheds provide a reliable source of water, reduce the risk of flooding, and improve water quality.

- **Wetlands Restoration:** Wetlands act as natural sponges, absorbing excess water during floods and releasing it during dry periods. Restoring and protecting wetlands can enhance water security, reduce flood risks, and support agricultural production. Wetlands also provide habitat for wildlife, improve water quality by filtering pollutants, and support biodiversity.
- **Efficient Irrigation Systems:** Implementing efficient irrigation systems, such as drip irrigation and rainwater harvesting, can optimize water use and reduce vulnerability to drought. Drip irrigation delivers water directly to the plant roots, minimizing evaporation and ensuring that crops receive adequate water. Rainwater harvesting involves collecting and storing rainwater for agricultural use, providing a reliable water source during dry periods. Sustainable water management practices ensure that crops receive sufficient water even during dry periods, enhancing resilience to water scarcity.

Biodiversity Conservation

Preserving genetic diversity in crops and livestock breeds enhances resilience by providing a broader base of traits that can withstand climate stresses. Biodiversity can buffer against environmental changes, pests, and diseases, ensuring stable agricultural production.

- **Crop Diversity:** Conserving traditional crop varieties that are adapted to local conditions can provide valuable genetic resources for breeding resilient crops. Diverse cropping systems can reduce the risk of total crop failure, as different crops may respond differently to climate stresses. Intercropping and polyculture, where multiple crops are grown together, can also enhance resilience by improving soil health, reducing pest pressure, and increasing overall productivity.

- **Livestock Diversity:** Preserving a wide range of livestock breeds can enhance resilience by providing animals with different traits that can thrive under various environmental conditions. Breeds that are resistant to specific diseases, heat-tolerant, or able to thrive on diverse forage types are valuable in maintaining productivity in changing climates. Breeding programmes and conservation efforts should prioritize maintaining genetic diversity in livestock populations.
- **Agro-forestry:** Integrating trees and shrubs into agricultural systems can provide multiple benefits, including improved soil health, increased biodiversity, and enhanced resilience to climate change. Agro-forestry systems can offer additional sources of income, such as fruit, nuts, and timber, contributing to economic stability. Trees in agro-forestry systems can also provide shade, reduce soil erosion, and improve water infiltration.
- **Community-Led Initiative:** Local communities can identify and prioritize adaptation strategies that are most relevant to their specific conditions and needs. By engaging in the planning process, communities can tailor solutions to their unique challenges, ensuring that they are practical and sustainable. Community-led initiatives can also foster social cohesion and collective action, strengthening community resilience.
- **Local Knowledge Integration:** Traditional farming practices, indigenous crop varieties, and local resource management techniques can provide valuable insights and solutions for climate adaptation. Recognizing and integrating this knowledge into adaptation planning can improve the relevance and sustainability of adaptation measures. For example, traditional water management systems, such as terraced fields and check dams, can enhance water retention and reduce soil erosion.

Capacity Building

Providing training and resources to farmers helps them adopt new practices and technologies. Effective extension services, farmer field schools, and demonstration projects can disseminate information about climate-smart agricultural practices, resilient crop varieties, and efficient resource management.

- **Extension Services:** Extension services play a crucial role in transferring knowledge and skills to farmers. These services can provide technical assistance, training programmes, and access to resources that enable farmers to adopt new practices. Extension agents can also facilitate the dissemination of research findings and innovations, helping farmers stay informed about the latest developments in agricultural adaptation.
- **Farmer Field Schools:** Farmer field schools offer hands-on training and experiential learning opportunities for farmers. Through field schools, farmers can learn about sustainable agricultural practices, pest management, and climate-resilient technologies. These schools promote peer-to-peer learning, encouraging farmers to share their experiences and solutions with each other.
- **Access to Resources:** Ensuring that farmers have access to the necessary resources, such as seeds, tools, and financial support, is critical for successful adaptation. Microfinance programmes, cooperative purchasing agreements, and subsidies can help farmers invest in new practices and technologies. Providing access to resources can also empower marginalized and resource-poor farmers, enhancing their resilience.

Conclusion

Improved resource management and community-based adaptation are essential for building resilient agricultural systems that can withstand the impacts of climate change. Sustainable management of soil, water, and biodiversity enhances agricultural productivity and resilience, while engaging local communities ensures that adaptation strategies are context-specific and culturally appropriate. By integrating local knowledge, fostering participatory planning, and providing capacity-building support, farmers can develop and implement effective adaptation measures. These strategies not only enhance agricultural productivity but also promote environmental sustainability and social well-being. Through concerted efforts and innovative solutions, it is possible to create resilient agricultural systems that can secure food supplies and support sustainable development in a changing climate.

Role of Technology and Innovation in Agricultural Adaptation

Technology and innovation play pivotal roles in enabling agricultural adaptation to climate change. Advances in biotechnology, digital tools, precision agriculture, and infrastructure development significantly enhance resilience and productivity, helping farmers cope with the increasingly adverse conditions brought about by climate change.

Biotechnology

Biotechnology is a critical tool for developing climate-resilient crop varieties. Through genetic modification and traditional breeding techniques, scientists can create crops that are better equipped to withstand extreme weather conditions and pest pressures. These advancements are essential for maintaining high yields and ensuring food security in vulnerable regions.

- **Climate-Resilient Crops:** Developing crops that can tolerate drought, heat, and pests is vital for adapting to changing environmental conditions. For example, genetically modified drought-tolerant maize can maintain productivity during prolonged dry periods, while flood-resistant rice can survive and yield well in inundated fields. These crops reduce the risk of crop failure and provide a stable food supply even under adverse conditions.

- **Nutrient Efficiency:** Biotechnology also enables the development of crops with improved nutrient uptake and use efficiency. Plants that can better utilize available nutrients require fewer inputs, such as fertilizers, reducing costs and environmental impacts. Enhanced nutrient efficiency is particularly important in areas with poor soil quality or limited access to fertilizers.
- **Disease Resistance:** Genetic modifications can confer resistance to various plant diseases, reducing the reliance on chemical pesticides and enhancing crop resilience. Disease-resistant crops can maintain yields even in the face of pathogen outbreaks, ensuring food security and reducing production losses.

Digital Tools

Digital technologies offer powerful tools for enhancing farm management and decision-making. Mobile applications, remote sensing, and data analytics provide real-time information and support informed agricultural practices.

- **Mobile Applications:** Mobile apps are invaluable for delivering critical information to farmers. These apps can provide weather forecasts, pest and disease alerts, and market price updates, helping farmers make timely decisions. For instance, an app that predicts rainfall can assist farmers in planning irrigation schedules, while pest alerts can prompt immediate action to protect crops.
- **Remote Sensing:** Satellite and drone-based remote sensing technologies enable the monitoring of large agricultural areas. These tools can detect crop health, soil moisture levels, and pest infestations from above, providing detailed insights that ground-based observations might miss. Remote sensing helps farmers respond quickly to emerging issues, optimizing resource use and improving crop management.

- **Data Analytics:** Advanced data analytics can process vast amounts of information to identify patterns and trends. By analyzing data on weather conditions, soil properties, and crop performance, farmers can make data-driven decisions that enhance productivity and resilience. Predictive analytics can forecast crop yields, guide planting decisions, and optimize input applications, reducing waste and maximizing efficiency.

Precision Agriculture

Precision agriculture leverages cutting-edge technologies to manage agricultural practices with high precision. This approach improves the efficiency of resource use and enhances productivity, making it a key strategy for adapting to climate variability.

- **Sensors:** Soil and crop sensors provide real-time data on soil moisture, nutrient levels, and crop health. This information allows farmers to apply water, fertilizers, and pesticides precisely where needed, minimizing waste and environmental impact. For example, moisture sensors can trigger irrigation systems only when soil moisture falls below a certain threshold, conserving water.
- **Drones:** Drones equipped with cameras and sensors can survey fields quickly and accurately. They can identify areas of stress, disease, or pest infestation, enabling targeted interventions. Drones can also assist in planting seeds, applying fertilizers, and monitoring crop growth, increasing efficiency and reducing labour costs.
- **GPS Technology:** GPS technology facilitates precise mapping and navigation in agricultural fields. Farmers can use GPS-guided equipment for planting, fertilizing, and harvesting, ensuring uniform application and reducing overlap and waste. This precision enhances overall field productivity and resource efficiency.

Infrastructure Development

Investing in infrastructure is essential for enhancing the resilience of agricultural systems. Improved infrastructure supports efficient production, reduces post-harvest losses, and ensures the timely delivery of inputs and produce.

- **Storage Facilities:** Adequate storage facilities are crucial for reducing post-harvest losses. Improved storage conditions, such as temperature and humidity control, can prevent spoilage and extend the shelf life of perishable goods. This is particularly important in regions where transportation delays are common or where markets are distant from production areas.
- **Transportation Networks:** Reliable transportation networks are vital for connecting farmers with markets and input suppliers. Good roads, railways, and ports facilitate the efficient movement of goods, reducing transportation costs and ensuring that produce reaches consumers quickly. Improved transportation infrastructure also enhances access to agricultural inputs, such as seeds, fertilizers, and equipment.
- **Irrigation Systems:** Efficient irrigation systems are critical for managing water resources in agriculture. Investments in infrastructure such as drip irrigation, sprinkler systems, and rainwater harvesting can optimize water use and ensure that crops receive adequate moisture even during dry periods. These systems reduce water wastage and increase the resilience of agricultural systems to drought.
- **Energy Infrastructure:** Reliable and affordable energy is essential for powering agricultural equipment, processing facilities, and storage units. Investments in renewable energy sources, such as solar and wind, can provide sustainable power solutions for rural areas. Solar-powered irrigation systems, for example, can reduce dependency on fossil fuels and lower operational costs.

Conclusion

Technology and innovation are crucial for enabling agricultural adaptation to climate change. Advances in biotechnology, digital tools, precision agriculture, and infrastructure development significantly enhance resilience and productivity, helping farmers cope with increasingly adverse conditions. By adopting these technologies and innovations, farmers can improve their agricultural practices, optimize resource use, and ensure food security in the face of climate change. Through continued investment and support for technological advancement, it is possible to create resilient agricultural systems that can sustain global food supplies and support sustainable development in a changing climate. Climate change poses significant challenges to agriculture and food security, with profound economic implications. To address these challenges, a combination of adaptation strategies, improved resource management, community-based approaches, and technological innovations is essential. By adopting climate-smart agricultural practices, leveraging technology, and fostering community resilience, it is possible to mitigate the impacts of climate change and ensure a sustainable and secure food supply. The integration of these strategies will not only enhance the resilience of agricultural systems but also promote economic stability and growth in a warming world.

REFERENCES

Auerbach, A.J., and Gorodnichenko, Y. (2022). "Fiscal Multipliers in Recessions and Expansions: Does It Matter Whether Government Spending is Increasing or Decreasing?" *Journal of Economic Perspectives, 36*(1): 41-64.

Ball, L.M., Gagnon, E., Honohan, P. and Krogstrup, S. (2022). "What Else Can Central Banks Do?" *IMF Economic Review, 70*(1): 167-206.

Christiano, L.J., Eichenbaum, M.S. and Trabandt, M. (2022). "Understanding the Great Inflation: Lessons for Current Policy". *Quarterly Journal of Economics, 137*(3): 1263-1300.

Clarida, R.H. (2022). "Monetary Policy, Price Stability, and Inflation: Theoretical Perspectives and Practical Lessons". *Journal of Economic Literature, 60*(1): 34-62.

Eggertsson, G.B., Mehrotra, N.R. and Robbins, J.A. (2022). "A Model of Secular Stagnation: Theory and Quantitative Evaluation". *American Economic Journal: Macroeconomics, 14*(1): 259-291.

Eichenbaum, M.S., Rebelo, S. and Wong, A. (2022). "State-Dependent Effects of Monetary Policy: The Asymmetric Role of Supply Constraints". *Review of Economic Studies, 89*(4): 721-61.

Guerrieri, V., Lorenzoni, G., Straub, L. and Werning, I. (2022). "Macroeconomic Implications of COVID-19: Can Negative Supply Shocks Cause Demand Shortages?" *American Economic Review: Insights, 4*(2): 173-190.

Ha, J., Kose, M.A. and Ohnsorge, F. (2022). "Global Inflation Synchronization". *Journal of International Economics, 134*, p. 103595.

Hall, R.E. (2022). "The Limits of Inflation Targeting". *Journal of Economic Perspectives, 36*(2): 3-24.

Guvenen, F., Kaplan, G. and Song, J. (2022). "How Does Inflation Affect Different Income Groups?" *Journal of Political Economy, 130*(5): 1437-1473.

Hamilton, J.D. (2022). "Why You Should Never Use the Hodrick-Prescott Filter". *Review of Economic Studies, 89*(1): 104-120.

Jordà, Ò. and Taylor, A.M. (2022). "The Pandemic, Inflation, and Debt Sustainability". *Economic Policy, 37*(110): 275-314.

Jordà, Ò., Singh, S.R. and Taylor, A.M. (2022). "The Long-Run Effects of Monetary Policy". *Journal of Monetary Economics, 123*: 1-20.

Kiley, M.T. (2022). "Unemployment, Inflation, and Monetary Policy: Evidence from Advanced and Emerging Market Economies". *Journal of International Money and Finance*, Vol. 122, Editorial Board Article No. 102564.

Krugman, P. (2022). "Inflation and the Power of Economic Ideas". *Foreign Affairs, 101*(2): 23-37.

Laubach, T. and Williams, J.C. (2022). "Measuring the Natural Rate of Interest: International Trends and Determinants". *Brookings Papers on Economic Activity, 2022*(3): 57-108.

Mishkin, F.S. (2022). "Monetary Policy Strategy: Lessons from the Financial Crisis". *Journal of Economic Perspectives, 36*(3): 123-148.

Mussa, M. (2022). "Inflation, Real Interest Rates, and the Real Exchange Rate: An Empirical Investigation". *Journal of Monetary Economics, 127*: 243-271.

Orphanides, A. (2022). "Policy Rules and Economic Stability: Lessons from the Great Inflation". *Journal of Economic Dynamics and Control,* Vol. 135, Editorial Board Article No. 104313.

Reis, R. (2022). "The Analytics of Monetary Policy: A New Keynesian Perspective". *Quarterly Journal of Economics, 137*(4): 1951-1987.

Sims, C.A. (2022). "Fiscal Policy, Inflation, and Debt Sustainability". *Journal of Economic Theory, 198.*

Smets, F. and Wouters, R. (2022). "An Estimated Dynamic Stochastic General Equilibrium Model of the Euro Area". *Journal of Economic Dynamics and Control, 133.*

Stock, J.H. and Watson, M.W. (2022). "Disentangling the Channels of the 2021-2022 Inflation Surge". *Journal of Applied Econometrics, 37*(6): 975-996.

7

Energy Transition Economics

Shifting Towards Renewable Resources

Economic Benefits of Renewable Energy

The transition to renewable energy sources offers numerous economic benefits, including job creation, energy security, and environmental sustainability. As the world shifts away from fossil fuels, renewable energy technologies such as solar, wind, hydro, and bio-energy are increasingly becoming cost-competitive and essential for sustainable economic growth.

- **Job Creation:** The renewable energy sector is a significant source of employment. The manufacturing, installation, and maintenance of renewable energy systems create jobs across various skill levels. For instance, the solar industry employs technicians, engineers, sales representatives, and maintenance workers. According to the International Renewable Energy Agency (IRENA), the renewable energy sector employed over 11 million people globally in 2018, with numbers expected to grow as the sector expands. This job creation can stimulate local economies, reduce unemployment, and foster economic development. The diversity of roles in the sector means that job opportunities are available in both urban and rural areas, providing a balanced economic uplift.

- **Energy Security:** Renewable energy sources enhance energy security by reducing dependence on imported fossil fuels. Countries can leverage their local renewable resources, such as solar, wind, and hydro, to produce energy domestically. This reduces vulnerability to global energy market fluctuations and geopolitical tensions. Increased energy independence can also lead to more stable energy prices, benefiting consumers and businesses alike. For instance, countries that invest in renewable energy infrastructure can buffer themselves against the volatile prices of oil and gas, ensuring a more predictable and stable energy economy.
- **Environmental Sustainability:** Renewable energy technologies produce little to no greenhouse gas emissions during operation, contributing significantly to efforts to combat climate change. By replacing fossil fuels with renewables, countries can significantly reduce their carbon footprints, improving air quality and public health.

Additionally, renewable energy systems often have lower environmental impacts compared to traditional energy sources. For example, solar and wind farms require less water and produce less waste than coal-fired power plants. This reduction in environmental harm translates into long-term economic benefits, such as reduced healthcare costs and the preservation of ecosystems that are vital for industries like agriculture and tourism.

- **Economic Diversification:** Investing in renewable energy can diversify economies by reducing reliance on finite fossil fuel resources. This diversification can create new industries and opportunities for innovation. As countries develop expertise in renewable energy technologies, they can export knowledge, products, and services, fostering international trade and economic growth. For example, Denmark's success in wind energy technology has allowed it to become a global leader in wind turbine manufacturing, exporting technology and expertise worldwide. This not only strengthens Denmark's economy but also builds global partnerships and markets for renewable energy.

- **Cost-Effectiveness:** The cost of renewable energy technologies has decreased significantly over the years, making them increasingly competitive with traditional fossil fuels. Advances in technology and economies of scale have driven down the prices of solar panels, wind turbines, and other renewable energy components. This cost reduction makes it more economically feasible for countries and businesses to invest in renewable energy. Furthermore, renewable energy systems have lowered operational and maintenance costs compared to fossil fuel plants, which can be expensive to run and maintain due to fuel costs and regulatory compliance. Over time, the savings from lower operational costs and the elimination of fuel expenses can make renewable energy a more attractive investment.
- **Resilience and Adaptability:** Renewable energy systems can enhance the resilience and adaptability of energy infrastructure. Decentralized renewable energy sources, such as rooftop solar panels and small-scale wind turbines, can provide energy security in remote and underserved areas, reducing the vulnerability of centralized power grids to disruptions. This decentralization can also support disaster recovery efforts by providing a reliable and independent power source in the aftermath of natural disasters.

  For instance, micro grids powered by renewable energy can keep critical services running when the main grid fails, ensuring that hospitals, emergency services, and communication networks remain operational.

- **Innovation and Technological Advancement:** The push towards renewable energy drives innovation and technological advancement. Research and development in renewable energy technologies can lead to breakthroughs that improve efficiency, reduce costs, and open up new applications. These innovations can have spillover effects into other sectors, such as electric vehicles, energy storage, and smart grid technologies, further stimulating economic growth. Governments and private sectors investing in renewable energy research create a culture of innovation that can position a country as a leader in the global clean energy market.

Conclusion

In conclusion, the economic benefits of renewable energy are substantial and multifaceted. From job creation and energy security to environmental sustainability and economic diversification, the transition to renewable energy offers a pathway to a more stable, prosperous, and sustainable future. By continuing to invest in renewable energy technologies and creating supportive policy frameworks, countries can harness these benefits to drive long-term economic growth and development.

Job Creation

The growth of the renewable energy sector generates employment opportunities that are diverse and widely distributed. Unlike the fossil fuel industry, which tends to concentrate jobs in specific regions with abundant resources, renewable energy jobs are more geographically dispersed. Solar panel installations can occur on rooftops in urban areas, wind turbines can be erected in rural and coastal regions, and bio-energy projects can be established near agricultural zones. This geographic spread ensures that a broader segment of the population can benefit from job opportunities in the renewable energy sector, leading to more balanced regional development.

Moreover, the renewable energy sector offers a range of job types, from highly skilled engineering and technical positions to entry-level roles that require minimal training. For instance, the production of wind turbines involves engineers and machinists, while their installation and maintenance require construction workers and technicians. Solar farms and wind parks also need ongoing maintenance and monitoring, providing long-term employment opportunities. This variety of jobs means that individuals with different educational backgrounds and skill sets can find employment in the renewable energy sector.

Making it an inclusive source of economic growth. In addition to direct job creation, the renewable energy sector stimulates employment in related industries. The demand for raw materials such as steel, concrete, and rare earth elements used in renewable energy technologies can boost employment in mining and manufacturing sectors. Furthermore, the construction of renewable energy projects requires services from the logistics, transportation, and legal sectors, creating a multiplier effect that benefits the broader economy. For example, the development of a large-scale wind farm requires extensive logistical support for transporting turbine components, legal expertise for land acquisition and regulatory compliance, and financial services for project financing.

The renewable energy sector also drives innovation and entrepreneurship, creating new business opportunities and job roles. Start-ups and small businesses play a crucial role in developing and deploying innovative renewable energy technologies and services. These companies often bring fresh ideas and agility to the market, driving competition and furthering technological advancements. Governments and private investors are increasingly recognizing the potential of these businesses, providing funding and support through incubators, accelerators, and grants. This entrepreneurial ecosystem fosters job creation and economic growth, particularly in regions that embrace innovation and sustainability. Furthermore, the transition to renewable energy aligns with global efforts to combat climate change, attracting significant public and private investment. Governments around the world are setting ambitious renewable energy targets and implementing supportive policies, such as subsidies, tax incentives, and renewable energy mandates.

These policies create a favourable environment for the renewable energy sector to flourish, encouraging investment and expansion. For instance, the European Union's Green Deal aims to make Europe the first climate-neutral continent by 2050, with a substantial portion of investments directed towards renewable energy projects. Such initiatives not only drive job creation within the renewable energy sector but also enhance the resilience and sustainability of the economy as a whole. Education and training programmes are crucial to preparing the workforce for jobs in the renewable energy sector. Many institutions are developing specialized curricula and certification programmes to equip individuals with the skills needed for careers in renewable energy. Technical schools and universities offer courses in renewable energy technology, engineering, and environmental science, while industry-specific training programmes provide hands-on experience with renewable energy systems.

For example, certification programmes for solar panel installers ensure that technicians have the necessary knowledge and skills to safely and efficiently install solar systems. These educational initiatives help bridge the skills gap and ensure a steady supply of qualified workers for growing renewable energy sector. The renewable energy sector also offers significant potential for improving gender diversity in the workforce. Traditionally, the energy sector has been male-dominated, but the growing renewable energy industry presents an opportunity to promote gender equality and inclusion. Women can and do play a critical role in the renewable energy sector, from research and development to project management and policy advocacy. Encouraging greater female participation through targeted recruitment, mentorship programmes, and inclusive workplace policies can enhance innovation and productivity within the sector.

By fostering a diverse and inclusive workforce, the renewable energy sector can leverage a wider range of talents and perspectives, driving economic growth and social progress.

Conclusion

In conclusion, the renewable energy sector is a powerful engine for job creation, offering diverse and widespread employment opportunities that stimulate local economies and reduce unemployment. By providing jobs across various skill levels and fostering innovation and entrepreneurship, the renewable energy sector contributes to sustainable economic development. Government support through favourable policies and investment, along with targeted education and training programmes, ensures that the workforce is prepared to meet the demands of this rapidly growing sector. Emphasizing diversity and inclusion further enhances the sector's potential, making renewable energy a key driver of both economic and social progress.

Energy Security

Renewable energy sources enhance energy security by reducing dependence on imported fossil fuels. Countries can leverage their local renewable resources, such as solar, wind, and hydro, to produce energy domestically. This shift diminishes vulnerability to global energy market fluctuations and geopolitical tensions. Increased energy independence can also lead to more stable energy prices, benefiting consumers and businesses alike.

1. **Reduction of Import Dependence:** One of the primary advantages of renewable energy is its potential to reduce a country's dependence on imported fossil fuels. Many nations rely heavily on oil, gas, and coal imports to meet their energy needs, exposing them to the risks of supply disruptions and price volatility. By investing in renewable energy infrastructure, countries can harness their abundant local resources, such as sunlight, wind, and water, to generate power.

This transition reduces the need for energy imports, enhancing national energy security. For example, solar and wind energy can be harnessed in vast areas, making it feasible for countries with varying geographic and climatic conditions to produce their own energy.

2. **Stability in Energy Prices:** Renewable energy can contribute to more stable energy prices. Fossil fuel markets are often characterized by significant price volatility due to geopolitical events, supply chain disruptions, and market speculation. In contrast, the costs of renewable energy sources, once the infrastructure is in place, are relatively stable and predictable. The sun shines, the wind blows, and rivers flow without the same degree of market fluctuations seen in fossil fuel markets. As a result, consumers and businesses can benefit from more predictable energy costs, which aid in financial planning and reduce the economic impact of energy price shocks.
3. **Mitigating Geopolitical Risks:** The global energy landscape is heavily influenced by geopolitical dynamics. Countries rich in fossil fuels can exert considerable influence on global markets, and political instability in these regions can lead to supply disruptions and price spikes. By investing in renewable energy, countries can insulate themselves from these geopolitical risks. For instance, Europe's focus on renewable energy has been partly driven by a desire to reduce dependence on Russian gas, enhancing energy security in the face of geopolitical uncertainties.
4. **Local Economic Development:** Investing in renewable energy not only enhances energy security but also stimulates local economic development. Building and maintaining renewable energy infrastructure creates jobs and generates economic activity in the regions where these projects are located. This can lead to increased local income and improved economic resilience. For example,

rural areas that host wind farms or solar arrays often see significant economic benefits from construction and ongoing maintenance activities, supporting local businesses and boosting employment.

5. **Resilience to Supply Chain Disruptions:** Renewable energy systems contribute to energy security by making energy supply chains more resilient. Unlike fossil fuels, which require complex and often international supply chains, renewable energy can be generated and used locally, reducing the risk of supply chain disruptions. For instance, a solar power plant does not rely on fuel shipments and is less vulnerable to transportation bottlenecks or geopolitical conflicts. This localized production and consumption model enhances the overall resilience of the energy system.
6. **Diversification of Energy Mix:** Diversifying the energy mix by incorporating renewable sources reduces the risk associated with dependence on a single type of energy. An energy system that relies on multiple sources—solar, wind, hydro, and biomass—is more robust against disruptions in any single source. For instance, during periods when solar power generation is low, wind or hydro power can compensate, ensuring a continuous and stable energy supply. This diversification is critical for maintaining energy security in the face of changing environmental and market conditions.
7. **Enhancement of National Security:** Energy security is intrinsically linked to national security. Countries that rely heavily on imported fossil fuels may find their foreign policy and national security strategies constrained by their energy dependencies. By developing renewable energy resources, nations can achieve greater autonomy and strategic flexibility. This shift allows countries to pursue foreign policy objectives without the encumbrance of energy dependency, thereby enhancing their overall security posture.

8. **Reduction in Infrastructure Vulnerability:** Renewable energy systems, particularly decentralized ones like rooftop solar panels and community wind farms, are less vulnerable to large-scale failures than centralized fossil fuel power plants. Decentralized energy production reduces the risk of massive outages due to natural disasters, accidents, or attacks on infrastructure. This distributed nature of renewable energy systems contributes to a more resilient energy grid, capable of maintaining operation under a wider range of conditions.
9. **Environmental and Health Benefits:** Enhancing energy security through renewable energy also brings significant environmental and health benefits. Reduced reliance on fossil fuels decreases greenhouse gas emissions, mitigating climate change and improving air quality. Better air quality leads to fewer health issues related to pollution, such as respiratory and cardiovascular diseases, reducing healthcare costs and improving public health. These environmental and health benefits contribute to the overall well-being and security of a nation.

Conclusion

In conclusion, renewable energy sources significantly enhance energy security by reducing dependence on imported fossil fuels, stabilizing energy prices, and mitigating geopolitical risks.

The local economic benefits, increased resilience to supply chain disruptions, diversification of the energy mix, and environmental advantages further underscore the importance of transitioning to renewable energy. By investing in renewable energy infrastructure, countries can achieve greater energy independence, economic stability, and national security, creating a more sustainable and secure future.

Environmental Sustainability

Renewable energy technologies produce little to no greenhouse gas emissions during operation, contributing significantly to efforts to combat climate change. By replacing fossil fuels with renewables, countries can markedly reduce their carbon footprints, thereby improving air quality and public health. Additionally, renewable energy systems often have lower environmental impacts compared to traditional energy sources. For example, solar and wind farms require less water and produce less waste than coal-fired power plants.

- **Reduction in Greenhouse Gas Emissions:** One of the most significant environmental benefits of renewable energy is its potential to drastically reduce greenhouse gas emissions. Fossil fuel combustion for electricity and heat is a major source of carbon dioxide (CO_2), methane (CH_4), and other greenhouse gases that contribute to global warming. Renewable energy sources like solar, wind, hydro, and geothermal produce electricity without emitting these gases. By transitioning to renewable energy, countries can decrease their greenhouse gas emissions, helping to mitigate the impacts of climate change. This shift is critical for meeting international climate targets, such as those outlined in the Paris Agreement, which aims to limit global temperature rise to well below 2 degrees Celsius above pre-industrial levels.
- **Improvement in Air Quality:** Burning fossil fuels not only emit greenhouse gases but also release pollutants such as sulphur dioxide (SO_2), nitrogen oxides (NO_x), and particulate matter (PM). These pollutants contribute to air quality problems, including smog and acid rain, and have severe health impacts. Renewable energy technologies, by contrast, generate power without emitting these harmful pollutants. For example, wind turbines and solar panels produce electricity without any combustion process, leading to zero emissions of SO_2, NO_x, and PM. Improved air quality resulting from the adoption of renewable energy can reduce the incidence of respiratory and cardiovascular

diseases, lower healthcare costs, and enhance overall public health.

- **Conservation of Water Resources:** Traditional fossil fuel power plants, especially coal and nuclear, consume vast amounts of water for cooling purposes. This can strain local water resources, particularly in arid regions or during periods of drought. In contrast, renewable energy technologies like wind and solar photovoltaic (PV) systems require minimal water for their operation. Hydropower, while water-intensive, generally involves the use of existing water flows without additional consumption. By reducing the water footprint of electricity generation, renewable energy can help conserve water resources, making them available for other critical uses such as agriculture, drinking water, and ecosystem support.
- **Reduction of Waste and Pollution:** Renewable energy systems typically produce less solid and hazardous waste than fossil fuel plants. Coal-fired power plants generate large quantities of ash and sludge, which can contain harmful heavy metals like mercury and lead. These wastes need to be managed and disposed of properly to avoid environmental contamination. Renewable energy technologies, on the other hand, produce minimal waste. For instance, solar panels and wind turbines generate no waste during operation, and even at the end of their life cycles, recycling and disposal of components are relatively straightforward and less harmful compared to fossil fuel wastes.
- **Biodiversity Protection:** The extraction and burning of fossil fuels can have devastating impacts on biodiversity. Oil spills, coal mining, and natural gas fracking destroy habitats, pollute land and water, and harm wildlife. Renewable energy projects, if properly sited and managed, can avoid many of these negative impacts. For example,

solar farms can be installed on rooftops, degraded lands, or deserts where they have minimal impact on wildlife. Wind farms can be located away from major bird migration paths to reduce collisions. Hydropower projects, if designed with ecological considerations, can maintain river ecosystems and fish populations. By reducing habitat destruction and pollution, renewable energy can play a crucial role in protecting biodiversity.

Mitigation of Climate Change Impacts

The adverse effects of climate change, such as more frequent and severe weather events, rising sea levels, and shifting ecosystems, pose significant risks to both natural and human systems. By lowering greenhouse gas emissions, renewable energy helps mitigate these impacts. This not only preserves the environment but also protects human communities from the economic and social disruptions caused by climate change.

For instance, reducing emissions can help stabilize weather patterns, thereby lessening the frequency and intensity of extreme weather events that can devastate infrastructure and agriculture.

- **Sustainable Development:** Renewable energy supports sustainable development by promoting long-term economic and environmental health. Fossil fuel resources are finite and will eventually be depleted, whereas renewable resources like sunlight and wind are abundant and inexhaustible. Investing in renewable energy infrastructure fosters economic growth and job creation without compromising environmental integrity. This aligns with the principles of sustainable development, which seek to meet the needs of the present without compromising the ability of future generations to meet their own needs.
- **Community and Ecosystem Benefits:** Renewable energy projects can also provide direct benefits to local communities and ecosystems. For example, community-owned renewable energy projects can generate income for

local residents and reduce energy costs. These projects can foster a sense of ownership and empowerment, encouraging sustainable practices and resilience. Additionally, renewable energy projects can be integrated with other land uses, such as agriculture or conservation, creating multifunctional landscapes that support both energy production and ecosystem services.

- **Reduction of Land Degradation:** Fossil fuel extraction often leads to significant land degradation, including soil erosion, deforestation, and contamination. Renewable energy projects, particularly solar and wind generally have a smaller land footprint and can be designed to minimize land disturbance. For instance, solar panels can be installed on degraded lands or integrated into buildings, reducing the need for additional land. Wind turbines occupy only a small portion of land, allowing for continued agricultural use or natural vegetation growth around them.

Conclusion

In conclusion, renewable energy technologies offer substantial environmental sustainability benefits. They significantly reduce greenhouse gas emissions, improve air quality, conserve water resources, reduce waste and pollution, protect biodiversity, mitigate climate change impacts, and support sustainable development. By transitioning to renewable energy, countries can achieve a cleaner, healthier, and more sustainable future, addressing both environmental and public health challenges while fostering economic growth.

Economic Diversification

Investing in renewable energy can diversify economies by reducing reliance on finite fossil fuel resources. This diversification can create new industries and opportunities for innovation. As countries develop expertise in renewable energy technologies, they can export knowledge, products, and services, fostering international trade and economic growth.

- **Reduction of Reliance on Fossil Fuels:** One of the primary benefits of economic diversification through renewable energy is the reduced dependence on fossil fuels, which are finite and subject to volatile market prices. By investing in renewable energy sources such as solar, wind, hydro, and geothermal, countries can decrease their vulnerability to fluctuations in global oil and gas prices, geopolitical tensions, and supply disruptions. This shift not only enhances energy security but also stabilizes the economy against external shocks associated with fossil fuel markets. Countries that are heavily dependent on fossil fuel exports can mitigate the economic risks posed by global market changes by developing a robust renewable energy sector.
- **Creation of New Industries:** The transition to renewable energy fosters the development of new industries and sectors. Manufacturing facilities for solar panels, wind turbines, batteries, and other renewable energy technologies create a multitude of jobs and business opportunities. For instance, the production of photovoltaic cells and wind turbine components requires specialized manufacturing processes, leading to the establishment of factories and supply chains dedicated to these technologies. This industrial growth generates employment across various levels, from skilled engineers and technicians to factory workers and logistics personnel, contributing to economic development and reducing unemployment rates.
- **Opportunities for Innovation:** The renewable energy sector is a hotbed for innovation. Continuous advancements in technology are necessary to improve the efficiency, cost-effectiveness, and scalability of renewable energy solutions. This innovation drives research and development (R&D) efforts in fields such as material science, electrical engineering, and data analytics. Universities, research institutions, and private companies invest in developing

cutting-edge technologies, fostering a culture of innovation that spills over into other sectors. For example, advances in battery storage technology not only benefit renewable energy systems but also have applications in electric vehicles, consumer electronics, and grid stability solutions.

- **Export of Knowledge and Technology:** As countries develop expertise in renewable energy technologies, they gain the ability to export knowledge, products, and services. Leading nations in renewable energy, such as Germany, Denmark, and China, have become global exporters of renewable energy technology and expertise. These countries provide technical assistance, equipment, and consulting services to other nations looking to develop their renewable energy sectors. This international trade fosters economic growth and strengthens geopolitical relationships, as countries collaborate on shared goals of sustainability and energy security.
- **Fostering International Trade:** The renewable energy sector opens up new avenues for international trade. Countries that excel in producing renewable energy technologies can export these products to global markets, driving economic growth and creating trade surpluses. For instance, China is a major exporter of solar panels and wind turbines, supplying equipments to numerous countries around the world. This trade not only benefits the exporting country by generating revenue but also helps importing countries accelerate their transition to renewable energy, fostering global cooperation in addressing climate change.
- **Economic Resilience:** Diversifying the energy mix with renewable sources enhances economic resilience. By reducing reliance on a single type of energy source, countries can better withstand economic downturns and energy crises. Renewable energy systems, which often involve decentralized and distributed generation, are less

susceptible to large-scale failures and disruptions. This resilience ensures a more stable and reliable energy supply, which is crucial for economic stability and growth. For example, during extreme weather events, decentralized solar and wind installations can continue to provide power even if centralized fossil fuel plants are compromised.

- **Development of Green Jobs:** The renewable energy sector is a significant source of green jobs, which contribute to sustainable economic growth. Green jobs encompass a wide range of roles, including research and development, manufacturing, installation, maintenance, and policy planning. These jobs not only support economic development but also promote environmental sustainability. Training and education programmes tailored to the renewable energy sector prepare the workforce for these emerging opportunities, ensuring that workers have the necessary skills to thrive in a green economy.
- **Enhancement of Energy Access:** Investing in renewable energy can improve energy access in remote and underserved areas. Renewable energy systems, such as off-grid solar panels and small-scale wind turbines, can provide electricity to communities that lack access to centralized power grids. This improvement in energy access supports economic development by enabling new business opportunities, improving healthcare and education, and enhancing the overall quality of life. For instance, rural areas in Africa and Asia have benefited from solar home systems that provide reliable electricity for lighting, cooking, and communication.
- **Environmental and Social Benefits:** The economic diversification achieved through renewable energy also brings substantial environmental and social benefits. By reducing greenhouse gas emissions, renewable energy mitigates climate change and its associated impacts, such as extreme weather events and sea-level rise.

This environmental sustainability supports long-term economic stability by protecting natural resources and ecosystems that are vital for agriculture, tourism, and other industries. Additionally, the social benefits of renewable energy, such as improved public health from reduced air pollution, contribute to a healthier and more productive workforce.

Conclusion

In conclusion, investing in renewable energy significantly diversifies economies by reducing reliance on finite fossil fuel resources, creating new industries, and fostering innovation. The development of expertise in renewable energy technologies enables countries to export knowledge, products, and services, promoting international trade and economic growth. Moreover, the transition to renewable energy enhances economic resilience, creates green jobs, improves energy access, and provides substantial environmental and social benefits. By embracing renewable energy, countries can achieve sustainable economic development and build a more resilient and prosperous future.

Challenges in Transitioning from Fossil Fuels

Despite the clear benefits, transitioning from fossil fuels to renewable energy presents several challenges. These include technological, financial, and infrastructural barriers that need to be addressed to achieve a successful energy transition.

Technological Challenges

- **Intermittency and Storage:** One of the primary technological challenges associated with renewable energy sources such as solar and wind is their intermittent nature. Solar power generation depends on sunlight, which varies throughout the day and is absent at night, while wind power depends on wind conditions, which can be unpredictable. To ensure a reliable energy supply, it is crucial to develop and deploy advanced energy storage solutions. Current battery technologies, such as lithium-ion batteries, are improving but still face challenges in terms of cost, capacity, and lifespan. Developing cost-effective, high-capacity storage solutions are essential for managing the intermittency of renewable energy and ensuring a stable energy supply.

- **Grid Integration:** Integrating renewable energy sources into existing power grids presents technical challenges. Traditional grids were designed for centralized power generation from fossil fuel plants, which provide a constant and controllable energy output. Renewable energy sources, by contrast, generate power in a decentralized manner and have variable outputs. This variability can cause fluctuations in the grid, requiring sophisticated grid management and balancing solutions. Upgrading grid infrastructure to accommodate renewable energy involves significant technical and financial investment, including the development of smart grids and advanced grid management systems.
- **Technological Maturity:** While some renewable energy technologies, such as solar PV and onshore wind, have reached a high level of maturity and cost-effectiveness, others are still in the developmental stage. Technologies like advanced bio fuels, offshore wind, and tidal energy require further research and development to become commercially viable. Investing in R&D to improve the efficiency, reduce the costs, and scale up these technologies is crucial for diversifying the renewable energy mix and addressing the unique energy needs of different regions.

Financial Barriers

- **High Initial Investment:** The transition to renewable energy requires substantial upfront capital investment. Building renewable energy infrastructure, such as solar farms, wind turbines, and energy storage systems, involves significant costs. While the operational and maintenance costs of renewable energy systems are generally lower than those of fossil fuel plants, the initial capital expenditure can be a barrier, especially for developing countries and small-scale projects. Securing financing for renewable energy projects is often challenging due to perceived risks and uncertainties associated with new technologies and markets.

- **Financing and Investment Risk:** Attracting investment in renewable energy projects can be difficult due to financial risks related to policy uncertainty, market volatility, and technological advancements. Investors may be hesitant to commit capital to projects that depend on evolving technologies or are subject to changing regulatory environments. Establishing stable and supportive policy frameworks, along with financial instruments such as green bonds and risk guarantees, can help mitigate these risks and encourage investment.
- **Subsidy and Policy Imbalance:** In many regions, fossil fuel industries continue to receive substantial subsidies, making it challenging for renewable energy to compete on a level playing field. Redirecting subsidies from fossil fuels to renewable energy projects can help bridge this gap, but such policy shifts can be politically challenging. Developing clear, consistent, and long-term policy support for renewable energy is essential to provide market confidence and drive investment.

Infrastructural Barriers

- **Grid Infrastructure:** Existing power grids were primarily designed for centralized, fossil fuel-based energy production. Transitioning to a decentralized renewable energy system requires significant upgrades to grid infrastructure. This includes expanding transmission and distribution networks to connect remote renewable energy projects, developing smart grids to manage variable energy inputs, and integrating energy storage solutions. These infrastructural upgrades are costly and require long-term planning and investment.

- **Land and Resource Use:** Renewable energy projects often require significant land and resources. For example, solar farms need large areas for panel installations, and wind turbines require space for optimal placement. Finding suitable land that does not conflict with agricultural use, conservation efforts, or residential areas can be challenging. Additionally, the extraction and processing of raw materials for renewable energy technologies, such as rare earth metals for wind turbines and lithium for batteries, have environmental and social impacts that need to be managed sustainably.
- **Supply Chain and Manufacturing Capacity:** The rapid scaling up of renewable energy deployment necessitates a robust supply chain and manufacturing capacity. This includes the production of solar panels, wind turbines, batteries, and other components.

Establishing and scaling up manufacturing facilities, securing raw materials, and developing efficient supply chains are critical for meeting the growing demand for renewable energy technologies. Disruptions in supply chains, such as those caused by geopolitical tensions or natural disasters, can pose significant challenges to the timely and cost-effective deployment of renewable energy projects.

Community and Social Acceptance

- **Public Perception and Acceptance:** Gaining public support for renewable energy projects is essential for their successful implementation. While renewable energy is generally viewed positively, specific projects can face opposition due to perceived impacts on landscapes, wildlife, and local communities. Engaging with communities, addressing concerns, and demonstrating the benefits of renewable energy projects are crucial for securing social acceptance. Public education campaigns and participatory planning processes can help build community support and ensure that projects are implemented in a socially responsible manner.

- **Employment Transition:** The shift from fossil fuels to renewable energy has significant implications for the workforce. Fossil fuel industries employ millions of workers worldwide, and the transition may lead to job losses in these sectors. Ensuring a just transition for affected workers involves providing retraining and reskilling programmes, social support, and job placement services. Governments and industries must work together to create new employment opportunities in the renewable energy sector and support workers through the transition process.

Conclusion

In conclusion, transitioning from fossil fuels to renewable energy presents several challenges, including technological, financial, and infrastructural barriers. Addressing these challenges requires coordinated efforts from governments, industries, and communities. Investment in R&D, supportive policy frameworks, financial instruments, and public engagement are essential for overcoming these barriers and achieving a successful energy transition. By addressing these challenges, countries can unlock the full potential of renewable energy, ensuring a sustainable and secure energy future.

Technological Challenges

While renewable energy technologies have advanced significantly, there are still technological hurdles to overcome. For example, the intermittent nature of solar and wind power requires efficient energy storage solutions to ensure a stable energy supply.

Current battery technologies are improving, but large-scale storage systems remain costly and require further development. Additionally, integrating renewable energy into existing grids poses technical challenges related to grid stability and management.

- **Financial Barriers:** The initial investment required for renewable energy projects can be substantial. While the cost of renewable technologies has decreased over the years, the upfront capital needed for installation and infrastructure development can be a barrier, especially in developing countries. Access to financing is crucial for the expansion of renewable energy, but many projects struggle to secure funding due to perceived risks and uncertainties.
- **Infrastructure Limitations:** Existing energy infrastructure is often designed around centralized fossil fuel power plants. Transitioning to decentralized renewable energy systems requires significant modifications to the grid. Upgrading transmission and distribution networks to accommodate variable renewable energy sources can be costly and time-consuming. Additionally, the siting of renewable energy projects can face opposition from local communities and environmental concerns, complicating the development process.
- **Policy and Regulatory Hurdles:** Effective policy and regulatory frameworks are essential for facilitating the energy transition. However, in many regions, policies and regulations are not yet fully supportive of renewable energy. Fossil fuel subsidies, lack of incentives for renewables, and bureaucratic hurdles can impede progress. Clear and consistent policies are needed to create a favourable environment for renewable energy investments.

Policy and Financial Incentives for Renewable Energy

To overcome the challenges associated with transitioning from fossil fuels to renewable energy, governments and financial institutions play a crucial role in supporting the transition through policy and financial incentives. These measures are essential for reducing financial barriers, encouraging investment, and fostering the development and deployment of renewable energy technologies.

Subsidies and Tax Incentives

Governments can provide subsidies and tax incentives to reduce the financial burden of renewable energy projects.

These incentives can take various forms, such as investment tax credits, production tax credits, and grants. For example, the U.S. Federal Investment Tax Credit (ITC) offers a tax credit for a percentage of the cost of installing solar energy systems, significantly reducing upfront costs for developers. By lowering the initial capital required, these incentives make renewable energy projects more financially viable and attractive to investors. Investment tax credits reduce the amount of income tax that an investor must pay, based on the investment made in renewable energy projects. Production tax credits provide financial benefits based on the amount of energy produced by the renewable energy systems. Grants and subsidies can cover a portion of the project costs, reducing the overall financial burden on developers and encouraging more investment in renewable energy infrastructure.

Feed-in Tariffs and Renewable Portfolio Standards

Feed-in Tariffs (FITs) guarantee a fixed price for renewable energy generated and fed into the grid. This policy provides a stable revenue stream for investors, encouraging the development of renewable energy projects. By ensuring that renewable energy producers receive a predictable and fair return on their investment, FITs help mitigate financial risks and attract more investments into the sector. Renewable Portfolio Standards (RPS) mandate that a certain percentage of electricity must come from renewable sources, driving demand and investment in the sector. RPS policies require utilities to purchase a specified amount of renewable energy, creating a market for renewable energy certificates (RECs). These certificates represent proof that electricity has been generated from renewable sources. Utilities can meet their RPS obligations by purchasing RECs, which incentivizes the generation of renewable energy and stimulates market growth.

Green Bonds and Climate Finance

Green bonds are financial instruments specifically earmarked for environmentally beneficial projects, including renewable energy. Issuing green bonds can attract investors interested in sustainable investments and provide the necessary capital for renewable energy projects. Green bonds offer a way for governments, corporations, and financial institutions to raise funds for projects that have positive environmental impacts, such as solar farms, wind parks, and energy-efficient buildings. Climate finance initiatives from international organizations, such as the Green Climate Fund (GCF), also support renewable energy development in developing countries. The GCF provides financial assistance to projects that help reduce greenhouse gas emissions and adapt to the impacts of climate change. By offering grants, loans, equity, and guarantees, the GCF mobilizes resources to support renewable energy projects in regions that may otherwise struggle to secure funding.

Research and Development (R&D) Funding

Governments and private sectors can invest in R&D to advance renewable energy technologies. Funding for R&D can accelerate the development of more efficient, cost-effective, and scalable renewable energy solutions. Public-Private Partnerships play a vital role in fostering innovation and bringing new technologies to market. R&D funding supports the exploration and development of new materials, designs, and processes that improve the performance and reduce the costs of renewable energy technologies. For example, research into advanced photovoltaic materials can lead to more efficient solar panels, while innovations in battery storage can enhance the reliability of energy systems. Government grants, tax incentives, and subsidies for R&D activities encourage private companies and research institutions to invest in renewable energy research, driving technological progress.

Public-Private Partnerships

Public-Private Partnerships (PPPs) are collaborations between government entities and private companies to fund, develop, and operate renewable energy projects. PPPs leverage the strengths of both sectors: the financial resources and regulatory support of governments, and the technical expertise and efficiency of private companies. These partnerships can help overcome financial and logistical barriers, accelerate project timelines, and ensure the successful implementation of renewable energy initiatives. Governments can provide land, regulatory support, and initial funding, while private companies bring technical know-how and operational management. By sharing risks and benefits, PPPs can attract more investment into the renewable energy sector and drive large-scale deployment of clean energy technologies.

Regulatory Frameworks and Standards

Establishing clear and consistent regulatory frameworks and standards is essential for promoting renewable energy. Governments can implement policies that streamline permitting processes, set technical standards, and create favourable market conditions for renewable energy projects. These frameworks provide certainty for investors and developers, reducing the bureaucratic hurdles that can delay or discourage renewable energy projects. Regulatory frameworks can also include measures to ensure grid access for renewable energy producers, such as priority dispatch and grid interconnection standards.

By guaranteeing that renewable energy can be efficiently integrated into the grid, these policies help maximize the use of clean energy sources and support the stability and reliability of the energy system. In conclusion, policy and financial incentives are critical for overcoming the challenges associated with transitioning to renewable energy. Subsidies, tax incentives, feed-in tariffs, renewable portfolio standards, green bonds, climate finance, and R&D funding all play essential roles in reducing financial barriers, encouraging investment, and fostering technological innovation.

Public-private partnerships and robust regulatory frameworks further support the development and deployment of renewable energy projects. By implementing these measures, governments and financial institutions can drive the growth of the renewable energy sector, ensuring a sustainable and secure energy future.

Case Studies on Successful Energy Transitions

Germany's Energiewende

Germany's Energiewende (energy transition) is one of the most ambitious efforts to shift towards renewable energy. The country has implemented comprehensive policies, including feed-in tariffs (FITs), renewable portfolio standards (RPS), and significant research and development (R&D) investments. As a result, Germany has seen a substantial increase in renewable energy capacity, with renewable accounting for over 40 per cent of electricity consumption in recent years. The transition has also stimulated job creation and technological innovation, positioning Germany as a leader in renewable energy.

Comprehensive Policy Framework

Germany's success with Energiewende can be largely attributed to its robust policy framework designed to support renewable energy development. FITs were introduced to guarantee fixed payments to renewable energy producers, ensuring a stable income and encouraging investment in renewable technologies. This policy reduced financial risks and made it attractive for both small-scale and large-scale investors to participate in the energy transition. RPS policies mandated that a certain percentage of electricity must come from renewable sources, creating a reliable market demand for renewable energy. This combination of FITs and RPS ensured that there was both a financial incentive to invest in renewable energy and a guaranteed market for the generated electricity.

Significant R&D Investments

Germany has also committed substantial resources to R&D in

renewable energy technologies. By funding research institutions and private sector innovation, the country has developed advanced renewable technologies and improved their cost-effectiveness. For instance, advancements in photovoltaic efficiency, wind turbine design, and energy storage systems have been pivotal in making renewable energy more competitive with traditional fossil fuels.

Job Creation and Economic Growth

The transition to renewable energy has had a positive impact on Germany's economy, particularly in terms of job creation. The renewable energy sector has generated thousands of jobs in manufacturing, installation, maintenance, and research. For example, the solar industry has created numerous employment opportunities in the production of solar panels, system installation, and ongoing system maintenance. The wind energy sector has similarly provided jobs in turbine manufacturing, logistics, and operational management. These jobs span a wide range of skills and expertise, from engineering and technical roles to sales and administrative positions, contributing to economic growth and reducing unemployment. The economic benefits extend beyond direct employment; the renewable energy sector has also stimulated growth in related industries, such as component manufacturing and supply chain logistics.

Technological Innovation and Export

Germany's investment in renewable energy has positioned it as a global leader in technological innovation. German companies are at the forefront of renewable energy technology development, producing some of the most efficient and reliable wind turbines, solar panels, and energy storage systems in the world. This technological leadership has allowed Germany to export renewable energy technologies and services, further boosting its economy and fostering international trade. The country's expertise in renewable energy has also led to the development of new business models and services, such as energy consulting and project management. German firms are now involved in renewable energy projects worldwide, providing technology, expertise, and management services to other countries looking to develop their renewable energy sectors.

Environmental and Social Benefits

The Energiewende has brought significant environmental benefits by reducing greenhouse gas emissions and air pollution. The increased use of renewable energy has decreased Germany's reliance on coal and natural gas, leading to lower carbon emissions and improved air quality. This transition supports Germany's commitments to international climate agreements and contributes to global efforts to combat climate change. Socially, the Energiewende has promoted greater public engagement and acceptance of renewable energy. Community-owned renewable energy projects have been a key feature of Germany's transition, allowing local communities to invest in and benefit from renewable energy generation. This model has increased public support for renewable energy and enhanced the sense of ownership and participation in the energy transition.

Challenges and Future Directions

Despite its successes, Germany's Energiewende has faced challenges, including the intermittency of renewable energy sources and the need for grid modernization. The country is addressing these issues through continued investment in energy storage technologies and smart grid infrastructure, aiming to enhance grid stability and integrate higher levels of renewable energy. Germany is also focusing on increasing energy efficiency and reducing overall energy consumption to complement its renewable energy efforts. Policies promoting energy-efficient buildings, transportation, and industrial processes are critical components of this strategy, ensuring a holistic approach to sustainable energy transition.

Conclusion

In conclusion, Germany's Energiewende exemplifies a successful energy transition driven by comprehensive policies, substantial R&D investments, and a commitment to renewable energy. The country's achievements in increasing renewable energy capacity, stimulating economic growth, and fostering technological innovation provide valuable lessons for other nations pursuing similar goals. By continuing to address challenges and invest in future technologies, Germany remains at the forefront of the global renewable energy movement, demonstrating the potential for a sustainable and prosperous energy future.

Costa Rica's Renewable Energy Leadership

Costa Rica is a global leader in renewable energy, generating nearly 100 per cent of its electricity from renewable sources, primarily hydro, wind, and geothermal. The government's commitment to environmental sustainability, supported by strong policy frameworks and investment in renewable infrastructure, has enabled this achievement. Costa Rica's success demonstrates the potential for small countries to achieve significant progress in renewable energy.

Government Commitment and Policy Frameworks

Costa Rica's remarkable success in renewable energy is rooted in its government's strong commitment to environmental sustainability. The country has set ambitious goals for renewable energy and implemented comprehensive policies to achieve them. Key among these is the National Energy Plan, which outlines strategies for increasing renewable energy capacity and reducing dependence on fossil fuels. The plan includes measures to promote energy efficiency, develop renewable energy projects, and enhance grid infrastructure. Costa Rica's policy framework includes incentives for renewable energy investments, such as tax exemptions, grants, and low-interest loans. These incentives have attracted both domestic and foreign investments in renewable energy projects, accelerating the country's transition to a sustainable energy system. Additionally, the government has established regulatory frameworks that facilitate the integration of renewable energy into the national grid, ensuring reliability and stability in electricity supply.

Investment in Renewable Infrastructure

Significant investment in renewable energy infrastructure has been crucial to Costa Rica's success. The country has developed extensive hydroelectric facilities, which account for the majority of its renewable energy generation. Hydropower is complemented by wind farms and geothermal plants, diversifying the energy mix and enhancing energy security. Costa Rica's investment in wind energy has been particularly noteworthy. The country has harnessed its favourable wind conditions to build several large-scale wind farms. These projects have not only increased renewable energy capacity but also created jobs and stimulated local economies.

Environmental and Economic Benefits

Costa Rica's renewable energy leadership has brought significant environmental and economic benefits. By generating nearly 100 per cent of its electricity from renewable sources, the country has drastically reduced its greenhouse gas emissions, contributing to global efforts to combat climate change. This achievement supports Costa Rica's commitment to becoming carbon-neutral by 2050, a goal that underscores its dedication to environmental stewardship. The transition to renewable energy has also improved air quality, with fewer pollutants emitted compared to fossil fuel-based power generation.

This has positive implications for public health, reducing the incidence of respiratory and cardiovascular diseases associated with air pollution. Economically, Costa Rica's renewable energy sector has created numerous jobs and fostered economic development. The construction, operation, and maintenance of renewable energy facilities require a skilled workforce, providing employment opportunities across various sectors. Additionally, the renewable energy industry has attracted investments, boosting local economies and enhancing the country's overall economic resilience.

Energy Independence and Security

One of the most significant advantages of Costa Rica's renewable energy transition is enhanced energy independence and security. By relying on domestically available renewable resources, Costa Rica has reduced its dependence on imported fossil fuels. This independence insulates the country from global oil price fluctuations and supply disruptions, ensuring a stable and reliable energy supply. The diversification of energy sources, including hydro, wind, and geothermal, has further strengthened Costa Rica's energy security. By tapping into multiple renewable resources, the country can maintain a consistent electricity supply even when one source is less productive due to seasonal variations or climatic conditions. This resilience is crucial for sustaining economic activities and supporting development goals.

Social and Community Engagement

Costa Rica's approach to renewable energy has been inclusive, involving local communities in decision-making processes and project implementation. Community engagement has been vital for gaining public support and ensuring the social acceptance of renewable energy projects.

Local communities have benefited from the economic opportunities created by renewable energy developments, such as job creation and infrastructure improvements. Programmes aimed at educating the public about the benefits of renewable energy and energy conservation have also played a significant role. By raising awareness and promoting sustainable practices, Costa Rica has fostered a culture of environmental responsibility and collective action towards sustainability goals.

Challenges and Future Prospects

Despite its successes, Costa Rica faces challenges in maintaining and expanding its renewable energy leadership. Climate variability, such as changes in rainfall patterns, can impact hydropower generation, necessitating further diversification of energy sources. The country is exploring additional renewable options, such as solar energy, to enhance its energy mix and increase resilience. Grid modernization and the integration of advanced technologies, such as smart grids and energy storage systems, are essential for optimizing renewable energy use and ensuring grid stability.

Continued investment in R&D and innovation will be crucial for addressing these challenges and driving future growth in the renewable energy sector. Costa Rica's success in renewable energy demonstrates that small countries can achieve significant progress with the right policies, investments, and community engagement. By prioritizing sustainability and leveraging its natural resources, Costa Rica has set an example for other nations aiming to transition to renewable energy and combat climate change.

Conclusion

In conclusion, Costa Rica's renewable energy leadership highlights the transformative potential of committed policies, substantial investments, and community involvement in achieving nearly 100 per cent renewable electricity generation. The country's experience offers valuable lessons for other nations seeking to develop sustainable energy systems, emphasizing the importance of comprehensive policy frameworks, diversified energy sources, and inclusive approaches to environmental sustainability.

China's Renewable Energy Expansion

China has rapidly expanded its renewable energy capacity, becoming the world's largest producer of solar and wind power. Government policies, including subsidies, feed-in tariffs (FITs), and substantial research and development (R&D) funding, have driven this growth.

- **China's Focus on Domestic Manufacturing of Renewable Energy Technologies:** This has also reduced costs and made renewable energy more competitive with fossil fuels. The country's commitment to reducing carbon emissions and improving air quality has further propelled its renewable energy transition.

- **Government Policies and Support:** The Chinese Government has played a pivotal role in the country's renewable energy expansion through robust policies and substantial financial support. Subsidies and FITs have provided significant financial incentives for renewable energy projects. FITs guarantee a fixed price for the electricity generated by renewable energy sources, ensuring a stable income for investors and encouraging the development of renewable energy infrastructure.
  These incentives have reduced the financial risks associated with renewable energy investments, making the sector more attractive to both domestic and international investors. Additionally, the Chinese Government has implemented Renewable Portfolio Standards (RPS) that require utilities to obtain a certain percentage of their electricity from renewable sources. This policy has created a reliable market for renewable energy, driving demand and stimulating investment in the sector. To further support the transition, the government has introduced various grants and low-interest loans aimed at reducing the initial capital costs of renewable energy projects.
- **Investment in R&D:** China's commitment to renewable energy is also reflected in its substantial investment in R&D. The country has allocated significant resources to research and development to improve the efficiency and cost-effectiveness of renewable energy technologies. By funding cutting-edge research in solar photovoltaics, wind turbine design, and battery storage, China has been able to achieve significant technological advancements. These improvements have not only reduced the costs of renewable energy technologies but have also enhanced their performance and reliability. The focus on R&D has

enabled China to develop domestic expertise in renewable energy technologies, fostering innovation and technological leadership. Chinese research institutions and companies are at the forefront of renewable energy innovation, developing new materials, designs, and processes that enhance the efficiency and reduce the costs of renewable energy systems.

- **Domestic Manufacturing and Cost Reduction:** One of the key factors behind China's success in renewable energy expansion is its emphasis on domestic manufacturing. By establishing a strong manufacturing base for renewable energy technologies, China has been able to produce solar panels, wind turbines, and other components at scale, significantly reducing production costs. Economies of scale, combined with technological advancements and government support, have made Chinese renewable energy products highly competitive in the global market. The cost reductions achieved through domestic manufacturing have been crucial in making renewable energy more affordable and accessible. Lower costs have encouraged the widespread adoption of renewable energy technologies, both within China and internationally. Chinese solar panels and wind turbines are now exported to numerous countries, contributing to global renewable energy capacity and helping other nations transition to sustainable energy systems.
- **Commitment to Carbon Emission Reduction:** China's renewable energy expansion is closely tied to its commitment to reducing carbon emissions and combating climate change. As the world's largest emitter of greenhouse gases, China has recognized the need to transition to cleaner energy sources to mitigate its environmental impact. The country has set ambitious targets for reducing carbon intensity and increasing the share of non-fossil fuels in its energy mix. The deployment of renewable energy

technologies has played a significant role in reducing China's reliance on coal and other fossil fuels. By increasing the share of renewable energy in its power generation mix, China has been able to decrease its carbon emissions and improve air quality. The reduction in emissions from coal-fired power plants has led to significant improvements in air quality, benefiting public health and reducing healthcare costs associated with pollution-related diseases.

- **Economic and Social Benefits:** The expansion of renewable energy in China has brought substantial economic and social benefits. The renewable energy sector has created millions of jobs across various segments, including manufacturing, installation, maintenance, and R&D. The growth of the renewable energy industry has stimulated local economies, particularly in regions where manufacturing facilities and renewable energy projects are located. Furthermore, the renewable energy transition has enhanced energy security by diversifying the energy mix and reducing dependence on imported fossil fuels. By harnessing its abundant renewable resources, China has increased its energy independence and resilience to global energy market fluctuations. This energy security is crucial for sustaining economic growth and stability.
- **Environmental Improvements:** The environmental benefits of China's renewable energy expansion are significant. The increased use of renewable energy sources has contributed to a substantial reduction in greenhouse gas emissions, supporting global efforts to mitigate climate change. Additionally, the reduction in coal consumption has led to improved air quality, reducing the incidence of respiratory and cardiovascular diseases caused by air pollution. China's renewable energy projects have also contributed to environmental conservation by reducing the ecological footprint associated with fossil fuel extraction and combustion. Hydropower, wind farms, and solar

installations have lower environmental impacts compared to coal and natural gas plants, helping to preserve natural habitats and biodiversity.

- **Challenges and Future Directions:** Despite its achievements, China faces challenges in continuing its renewable energy expansion. Integrating high levels of renewable energy into the grid requires significant investments in grid infrastructure and energy storage solutions. The intermittency of solar and wind power necessitates the development of advanced grid management systems and storage technologies to ensure a stable and reliable energy supply. Moreover, China must address the environmental and social impacts of its renewable energy projects. Large-scale hydropower projects, for example, can have significant ecological and social consequences, including habitat disruption and displacement of communities.

Ensuring that renewable energy projects are developed sustainably and responsibly is crucial for maintaining public support and achieving long-term environmental goals.

Conclusion

In conclusion, China's renewable energy expansion is a remarkable example of how government policies, substantial investments, and domestic manufacturing can drive a successful transition to renewable energy. The country's achievements in increasing renewable energy capacity, reducing costs, and improving air quality provide valuable lessons for other nations seeking to develop sustainable energy systems. By continuing to invest in R&D, enhance grid infrastructure, and address environmental and social challenges, China can maintain its leadership in the global renewable energy transition and contribute to a more sustainable and resilient future.

Denmark's Wind Energy Success

Denmark is a pioneer in wind energy, with wind power accounting for nearly half of the country's electricity consumption. Government policies, including long-term planning, investment in research and development (R&D), and supportive regulatory frameworks, have been instrumental in this success. Denmark's experience highlights the importance of consistent and forward-thinking policies in achieving renewable energy goals.

- **Long-term Planning and Vision:** Denmark's success in wind energy is rooted in its long-term planning and visionary approach to renewable energy. As early as the 1970s, Denmark recognized the potential of wind energy as a sustainable alternative to fossil fuels. This foresight led to the establishment of ambitious targets for renewable energy, with wind power playing a central role. The Danish Government set clear and progressive goals for wind energy development, providing a stable and predictable policy environment that encouraged investment and innovation. The Danish Energy Agreement, a series of long-term policy frameworks, has been crucial in maintaining the momentum of wind energy development. These agreements outline specific targets for renewable energy, investment plans, and policy measures to support the renewable energy sector. The continuity and clarity of these policies have provided confidence to investors, developers, and other stakeholders, facilitating sustained growth in the wind energy sector.
- **Investment in R&D and Innovation:** Investment in R&D has been a cornerstone of Denmark's wind energy success. The Danish Government, along with private sector partners, has invested heavily in research and technological development to enhance the efficiency, reliability, and cost-effectiveness of wind energy technologies. These investments have led to significant advancements in wind

turbine design, materials, and manufacturing processes. Denmark's commitment to R&D is exemplified by the establishment of institutions such as the Danish Wind Industry Association (DWIA) and the Danish Technical University (DTU), which have played pivotal roles in advancing wind energy technology. Collaborative efforts between academia, industry, and government have fostered innovation, leading to the development of state-of-the-art wind turbines that are among the most efficient and reliable in the world. The focus on innovation has also enabled Denmark to pioneer offshore wind energy.

The country is home to the world's first offshore wind farm, Vindeby, which was established in 1991. Since then, Denmark has continued to lead in offshore wind energy development, leveraging its expertise to build larger and more advanced offshore wind farms that contribute significantly to its renewable energy capacity.

- **Supportive Regulatory Frameworks:** Denmark's regulatory frameworks have been instrumental in creating a conducive environment for wind energy development. The government has implemented policies that ensure favourable conditions for wind energy projects, including streamlined permitting processes, grid access guarantees, and financial incentives.One of the key regulatory mechanisms that have supported wind energy growth is the feed-in tariff (FIT) system. FITs provide a guaranteed price for the electricity generated by wind turbines, ensuring a stable revenue stream for developers and investors. This financial certainty has encouraged the deployment of wind energy projects and facilitated the scaling up of wind power capacity. In addition to FITs, Denmark has implemented renewable energy certificates (RECs) and auction schemes that promote competition and cost reduction in the wind energy sector. These mechanisms have driven efficiency and innovation, making wind energy increasingly competitive with conventional energy sources.

- **Community and Stakeholder Engagement:** Denmark's wind energy success is also attributed to strong community and stakeholder engagement. The Danish model emphasizes local ownership and involvement in wind energy projects, which has fostered public support and minimized opposition. Many wind farms in Denmark are partially or fully owned by local cooperatives, farmers, and municipalities, ensuring that the economic benefits of wind energy are shared with local communities. This inclusive approach has built a sense of ownership and pride among Danish citizens, who are actively involved in the renewable energy transition. Public engagement campaigns and transparent decision-making processes have further strengthened community support for wind energy projects, contributing to their successful implementation.
- **Economic and Environmental Benefits:** The growth of wind energy in Denmark has brought substantial economic and environmental benefits. Economically, the wind energy sector has created thousands of jobs across various segments, including manufacturing, installation, maintenance, and R&D. Danish companies such as Vestas and Ørsted have become global leaders in wind energy technology and project development, exporting their expertise and products worldwide.

  The expansion of wind energy has also attracted significant foreign investment, boosting the Danish economy and enhancing its global competitiveness. The wind energy sector's contribution to GDP has increased steadily, reflecting its importance to Denmark's economic growth and development. Environmentally, the shift to wind energy has significantly reduced Denmark's greenhouse gas emissions and reliance on fossil fuels. Wind power generation produces no carbon emissions during operation, contributing to cleaner air and a reduction in climate change impacts. The transition to wind energy supports Denmark's commitment to international climate agreements and its goal of becoming carbon-neutral by 2050.

- **Challenges and Future Directions:** Despite its success, Denmark faces challenges in further expanding its wind energy capacity. Integrating high levels of wind power into the grid requires advanced grid management and energy storage solutions to ensure stability and reliability. The intermittency of wind energy necessitates investments in smart grid technologies and battery storage to balance supply and demand effectively. Additionally, Denmark must address environmental and social concerns related to wind energy development. Offshore wind farms, while less intrusive than onshore projects, can impact marine ecosystems and require careful planning and management. Ensuring that wind energy projects are developed sustainably and with minimal environmental impact is crucial for maintaining public support and achieving long-term renewable energy goals.

Conclusion

In conclusion, Denmark's wind energy success is a testament to the importance of consistent and forward-thinking policies, substantial investment in R&D, and supportive regulatory frameworks. The country's achievements in increasing wind power capacity, fostering innovation, and creating economic and environmental benefits provide valuable lessons for other nations pursuing renewable energy transitions. By continuing to address challenges and invest in future technologies, Denmark can maintain its leadership in the global wind energy sector and contribute to a sustainable and resilient energy future. The transition to renewable energy is essential for sustainable economic growth, energy security, and environmental sustainability. While the shift from fossil fuels presents challenges, such as technological, financial, and infrastructural barriers, effective policy and financial incentives can facilitate progress.

Case studies from Germany, Costa Rica, China, and Denmark demonstrate that with the right support, significant advances in renewable energy are achievable. By continuing to invest in renewable energy technologies and creating supportive policy frameworks, countries can secure a cleaner, more sustainable energy future.

REFERENCES

Aghion, P., Bergeaud, A., Cette, G. and Blundell, R. (2022). "The Impact of Inflation on Innovation: Evidence from Patent Data". *Journal of Monetary Economics, 131*: 77-92.

Bernanke, B.S. (2022). "The Real Effects of Disinflation". *Brookings Papers on Economic Activity, 2022*(1): 77-123.

Blinder, A.S. (2022). "Macroeconomics Under the Influence: Inflation, Expectations, and Policy". *Journal of Economic Perspectives, 36*(4): 65-88.

Guvenen, F., Kaplan, G. and Song, J. (2022). "How Does Inflation Affect Different Income Groups?" *Journal of Political Economy, 130*(5): 1437-1473.

Hamilton, J.D. (2022). "Why You Should Never Use the Hodrick-Prescott Filter", *Review of Economic Studies, 89*(1): 104-120.

Jordà, Ò. and Taylor, A.M. (2022). "The Pandemic, Inflation, and Debt Sustainability". *Economic Policy, 37*(110): 275-314.

Kamin, S.B. (2022). "Global Supply Chains in a Post-Pandemic World: Economic Insights and Policy Responses". *International Journal of Central Banking, 18*(3): 57-88.

Kiley, M.T. (2022). "Unemployment, Inflation, and Monetary Policy: Evidence from Advanced and Emerging Market Economies". *Journal of International Money and Finance*, Vol. 122, Editorial Board Article No. 102564.

Lane, P.R. (2022). "Global Supply Chains and Inflation: The Role of the Pandemic and Policy Responses". *IMF Economic Review, 70*(2): 121-143.

Laubach, T. and Williams, J.C. (2022). "Measuring the Natural Rate of Interest: International Trends and Determinants". *Brookings Papers on Economic Activity, 2022*(3): 57-108.

Martin, F. (2022). "Inflation, Wage Setting, and the Monetary Transmission Mechanism". *Journal of Economic Theory*, Vol. 197, Editorial Board Article No. 105299.

Mian, A., Sufi, A. and Verner, E. (2022). "Household Debt and Inflation: The Role of Mortgage Debt in Monetary Policy Transmission". *Journal of Financial Economics, 143*(3): 1041-1061.

Mishkin, F.S. (2022). "Monetary Policy Strategy: Lessons from the Financial Crisis". *Journal of Economic Perspectives, 36*(3): 123-148.

Mussa, M. (2022). "Inflation, Real Interest Rates, and the Real Exchange Rate: An Empirical Investigation". *Journal of Monetary Economics, 127*: 243-271.

Obstfeld, M. (2022). "Global Macroeconomic Challenges: Inflation, Supply Chains, and Policy Responses". *Brookings Papers on Economic Activity, 2022*(2): 37-74.

Orphanides, A. (2022). "Policy Rules and Economic Stability: Lessons from the Great Inflation". *Journal of Economic Dynamics and Control*, Vol. 135, Editorial Board Article No. 104313.

Reis, R. (2022). "The Analytics of Monetary Policy: A New Keynesian Perspective". *Quarterly Journal of Economics, 137*(4): 1951-1987.

Shiller, R.J. (2022). *Narrative Economics: How Stories Go Viral and Drive Major Economic Events*, 2nd edn., Princeton University Press.

Sims, C.A. (2022). "Fiscal Policy, Inflation, and Debt Sustainability". *Journal of Economic Theory*, Vol. 198, Editorial Board Article No. 105297.

Smets, F. and Wouters, R. (2022). "An Estimated Dynamic Stochastic General Equilibrium Model of the Euro Area". *Journal of Economic Dynamics and Control*, Vol. 133, Editorial Board Article No. 104295.

Stock, J.H. and Watson, M.W. (2022). "Disentangling the Channels of the 2021-2022 Inflation Surge". *Journal of Applied Econometrics, 37*(6): 975-996.

Woodford, M. (2022). "Inflation and Financial Stability: A Microeconomic Perspective". *Journal of Economic Literature, 60*(2): 337-365.

8

Coastal Cities and Economic Adaptation

Mitigating Risks from Rising Seas

Economic Risks for Coastal Cities

Coastal cities face significant economic risks from rising sea levels, driven by climate change. These risks include damage to infrastructure, loss of property, disruption of economic activities, and increased costs for emergency response and recovery.

- **Damage to Infrastructure:** Rising Sea levels threaten critical infrastructure such as roads, bridges, ports, and airports. Flooding and storm surges can cause extensive damage to these facilities, leading to costly repairs and disrupting transportation and commerce. For example, major coastal cities like New York, Miami, and Tokyo have experienced significant infrastructure damage due to coastal flooding and storms. This damage not only incurs direct repair costs but also results in economic losses from disrupted supply chains, delayed transportation, and hindered access to essential services.

- **Loss of Property:** Coastal cities are densely populated and have high property values. Sea-level rise and increased flooding can lead to the loss of residential and commercial properties, reducing property values and leading to substantial economic losses for property owners and local governments. Properties in flood-prone areas are particularly vulnerable, and the depreciation in property values can have a cascading effect on the local economy, reducing tax revenues and increasing the financial burden on municipal budgets. Additionally, insurance costs for properties in these areas are likely to rise, further increasing the financial strain on homeowners and businesses.
- **Disruption of Economic Activities:** Coastal cities are often economic hubs with diverse industries, including tourism, shipping, finance, and manufacturing. Flooding and storm surges can disrupt these economic activities, leading to loss of revenue and productivity. For example, tourism-dependent cities like Venice and Miami face significant risks from sea-level rise, which can damage beaches, hotels, and other tourist attractions. Ports and shipping lanes are also at risk, potentially disrupting global trade routes and economic activities. The financial sector, with its concentration of offices in coastal cities, could face significant operational disruptions, impacting global markets.
- **Increased Costs for Emergency Response and Recovery:** The frequency and intensity of coastal flooding and storms necessitate substantial investments in emergency response and recovery efforts. Local governments must allocate resources for disaster preparedness, evacuation plans, and post-disaster rebuilding. These costs can strain municipal budgets and divert funds from other essential services. For instance, the cost of emergency measures and rebuilding efforts after events like Hurricane Sandy in New York City runs into billions of dollars, putting immense pressure on

public finances. This scenario is likely to be repeated in other coastal cities facing similar threats, underscoring the need for proactive investment in resilience measures.

- **Impact on Infrastructure Investments:** The risks posed by rising sea levels can also deter infrastructure investments in coastal cities. Investors may be wary of funding new projects in areas vulnerable to flooding and erosion, which can stymie economic development and urban renewal initiatives. This hesitancy can slow down critical infrastructure upgrades and expansions necessary for supporting growing urban populations and economic activities. For instance, the potential for future flooding may discourage investments in new housing developments, commercial buildings, and transportation networks in affected areas, leading to stagnation in urban growth and economic progress.
- **Insurance and Financing Challenges:** As the risk of flooding and sea-level rise increases, insurance companies may hike premiums or withdraw coverage altogether for properties in high-risk areas. This shift can make it financially unfeasible for property owners to maintain insurance, leaving them vulnerable to losses. Furthermore, the rising cost of insurance can deter new investments, as potential developers and investors factor in the higher risk premiums into their cost calculations. Financing for coastal infrastructure projects may become more challenging to secure, requiring higher interest rates or additional guarantees, which can further slow down development and increase the overall cost of projects.
- **Social and Community Impact:** Beyond the direct economic risks, rising sea levels also pose significant social challenges. Displacement due to flooding can lead to the loss of homes and community structures, creating long-term social and economic instability. Vulnerable populations, including low-income households and marginalized communities, are often the most affected, exacerbating existing inequalities.

The disruption of social services, healthcare, and education due to flooding can have profound impacts on community well-being and long-term economic prospects. Addressing these social dimensions is crucial for creating resilient and inclusive coastal cities.

Conclusion

In conclusion, coastal cities face a multitude of economic risks from rising sea levels, including infrastructure damage, property loss, disrupted economic activities, increased emergency response costs, challenges to infrastructure investments, and insurance and financing difficulties. Proactive measures and comprehensive planning are essential to mitigate these risks and ensure the long-term resilience and economic sustainability of coastal cities. Investing in adaptive infrastructure, implementing robust land-use planning, and securing diverse funding mechanisms are critical steps towards safeguarding these urban centres from the impending impacts of climate change.

Cost Implications and Funding Mechanisms

The implementation of coastal adaptation strategies involves significant costs, but various funding mechanisms can support these efforts.

- **Cost Implications:** Building and maintaining physical defences, implementing land-use planning measures, and investing in natural and hybrid solutions require substantial financial resources. The costs of adaptation must be weighed against the potential economic losses from sea-level rise. While upfront investment in adaptation measures can be high, the long-term savings from reduced damage and disruption can justify the expenditure.
- **Funding Mechanisms:** Coastal cities can explore several funding mechanisms to finance adaptation projects. These mechanisms include:

- **Government Funding:** Local, state, and national governments can allocate budgetary resources for coastal adaptation projects. Governments can also apply for grants and funding from international organizations and climate funds, such as the Green Climate Fund (GCF).
- **Public-Private Partnerships (PPPs):** PPPs can leverage private sector investment and expertise to fund and implement adaptation projects. For example, private companies can invest in the construction and maintenance of seawalls or floodgates in exchange for concessions or revenue-sharing agreements.
- **Insurance and Risk Transfer Mechanisms:** Innovative insurance products, such as parametric insurance and catastrophe bonds, can provide financial protection against the impacts of sea-level rise. These mechanisms transfer risk from coastal cities to insurance markets, providing quick payouts for recovery and rebuilding efforts.
- **Climate Bonds and Green Financing:** Issuing climate bonds and securing green financing can raise capital for large-scale adaptation projects. These financial instruments attract investors interested in supporting sustainable and resilient infrastructure development.

Case Studies of Coastal Adaptation Projects

- **Rotterdam, Netherlands:** Rotterdam is a leading example of innovative coastal adaptation. The city has implemented a comprehensive adaptation strategy that includes the construction of advanced flood defences, such as the Maeslantkering storm surge barrier, and the development of "water plazas" that temporarily store rainwater during heavy rainfall. Rotterdam's approach combines engineering solutions with sustainable urban planning, creating a resilient and livable city.
- **Venice, Italy:** Venice faces unique challenges due to its location and historical significance. The MOSE (Modulo Sperimentale Elettromeccanico) project involves a system

of mobile barriers designed to protect the city from high tides and flooding. When not in use, the barriers lie on the seabed, but they can be raised to block incoming tides. The project aims to preserve Venice's cultural heritage while protecting its residents and economy.

- **Jakarta, Indonesia:** Jakarta is addressing its severe flooding problems with a combination of measures, including the construction of seawalls, river dredging, and the relocation of vulnerable communities. The city's "Giant Sea Wall" project aims to protect Jakarta from rising sea levels while creating new urban spaces through land reclamation. This ambitious plan seeks to balance economic development with environmental sustainability.

Conclusion

Coastal cities face significant economic risks from rising sea levels, but they can employ various strategies to mitigate these risks. Building physical defences, implementing land-use planning and zoning regulations, and investing in natural and hybrid solutions are crucial for protecting coastal infrastructure, properties, and economies. While the cost implications of these measures are substantial, diverse funding mechanisms, including government funding, PPPs, insurance products, and climate bonds, can support adaptation efforts. Case studies from Rotterdam, New York City, Venice, and Jakarta demonstrate that innovative and proactive approaches can enhance the resilience of coastal cities, safeguarding them against the impacts of climate change. By learning from these examples and investing in comprehensive adaptation strategies, coastal cities can mitigate the risks of rising seas and secure a sustainable future.

Strategies for Mitigating Sea-Level Rise

Coastal cities can employ various strategies to mitigate the risks associated with rising sea levels. These strategies include building physical defences, implementing land-use planning and zoning regulations, and investing in natural and hybrid solutions.

- **Building Physical Defences:** Constructing physical barriers such as seawalls, levees, and floodgates can protect coastal cities from storm surges and flooding. For example, the Thames Barrier in London and the Maeslantkering in Rotterdam are significant engineering projects designed to protect these cities from rising sea levels. These structures can provide robust protection but require substantial investment and maintenance. Seawalls and levees act as barriers that prevent water from inundating coastal areas, while floodgates can be opened or closed to control water flow during high tide or storm surge events. Although these defences are effective, they must be regularly maintained and upgraded to cope with increasing sea levels and storm intensity, ensuring they continue to provide adequate protection over time.
- **Implementing Land-use Planning and Zoning Regulations:** Effective land-use planning and zoning regulations can reduce vulnerability to sea-level rise by restricting development in high-risk areas. Coastal cities can implement policies that discourage construction in flood-prone zones and encourage the development of green spaces and natural buffers. These measures can reduce the potential for property damage and loss of life during flooding events. For instance, cities can designate certain areas as no-build zones to prevent new construction in regions that are highly susceptible to flooding.

  In addition, they can promote the development of parks, wetlands, and other green spaces that can absorb floodwaters and provide a buffer against rising seas. By integrating sea-level rise projections into urban planning, cities can ensure that future development is resilient and sustainable.
- **Investing in Natural and Hybrid Solutions:** Natural and hybrid solutions, such as restoring wetlands, mangroves, and coral reefs, can provide cost-effective protection

against sea-level rise. These ecosystems act as natural barriers, absorbing storm surges and reducing the impact of flooding. For example, wetlands can absorb excess water during high tide and heavy rainfall, while mangroves and coral reefs can dissipate wave energy, reducing the force of storm surges. Hybrid solutions combine natural and engineered approaches, such as using vegetation to stabilize constructed levees. These strategies not only protect coastal cities but also provide ecological benefits, enhancing biodiversity and supporting fisheries. Restoring and preserving natural ecosystems can be a sustainable way to mitigate the impacts of sea-level rise while promoting environmental conservation.

- **Retrofitting Existing Infrastructure:** In addition to building new defences, retrofitting existing infrastructure can enhance resilience to sea-level rise. This includes elevating buildings, roads, and utilities, and reinforcing structures to withstand flooding and storm surges. Cities can invest in elevating critical infrastructure such as power stations, wastewater treatment plants, and transportation networks to prevent disruption during flood events. Retrofitting also involves improving drainage systems to handle increased rainfall and tidal surges, ensuring that urban areas remain functional and accessible during extreme weather.
- **Community-based Adaptation:** Engaging local communities in adaptation planning and implementation is crucial for the success of mitigation strategies. Community-based adaptation involves educating residents about the risks of sea-level rise and involving them in decision-making processes. This approach ensures that adaptation measures are context-specific and culturally appropriate, gaining public support and increasing the likelihood of successful implementation. Communities can participate in activities such as creating flood action plans, maintaining natural

buffers, and advocating for resilient urban development. By empowering communities, cities can foster a sense of ownership and responsibility for adaptation efforts.

- **Monitoring and Early Warning Systems:** Implementing monitoring and early warning systems can enhance preparedness and response to sea-level rise and related hazards. These systems involve real-time monitoring of sea levels, weather patterns, and tidal conditions, providing timely information to authorities and residents. Early warning systems can trigger alerts for impending floods or storms, allowing for timely evacuations and protective measures. Effective monitoring and communication are essential for minimizing the impacts of sea-level rise and ensuring the safety of coastal population.
- **Financial Mechanisms and Incentives:** Securing funding for adaptation projects is critical for their implementation and sustainability. Coastal cities can explore various financial mechanisms and incentives to support adaptation efforts. This includes government funding, public-private partnerships, and innovative financing options such as climate bonds and insurance products. Providing financial incentives for property owners to implement flood-proofing measures and investing in resilient infrastructure can encourage widespread adoption of adaptation practices. Additionally, international grants and funds can support large-scale projects, especially in developing regions with limited financial resources.

Conclusion

In conclusion, coastal cities can employ a combination of strategies to mitigate the risks associated with rising sea levels. Building physical defences, implementing effective land-use planning and zoning regulations, and investing in natural and hybrid solutions are key approaches to enhancing resilience. Retrofitting existing infrastructure, engaging communities in adaptation efforts, and establishing monitoring and early warning systems further strengthen these strategies. By securing diverse funding mechanisms and incentives, coastal cities can ensure the successful implementation and sustainability of adaptation measures, protecting their population and economies from the impacts of climate change.

Investing in Natural and Hybrid Solutions

Natural and hybrid solutions offer a sustainable and cost-effective approach to protecting coastal cities from the impacts of sea-level rise. These strategies leverage the inherent resilience of natural ecosystems to mitigate flooding and storm surges while providing additional ecological benefits.

Restoring Wetlands

Wetlands serve as natural sponges, absorbing excess water during heavy rains and high tides. By restoring wetlands, coastal cities can create natural barriers that reduce the velocity and volume of floodwaters. Wetlands also filter pollutants, improving water quality and providing critical habitats for various species. Restoration projects can involve replanting native vegetation, reintroducing wetland hydrology, and removing invasive species. The restoration of wetlands not only helps in flood mitigation but also supports biodiversity, recreational activities, and carbon sequestration, contributing to broader environmental and social goals.

Rehabilitating Mangroves

Mangroves are highly effective in protecting shorelines from erosion and storm surges. Their dense root systems stabilize sediments and reduce the energy of incoming waves, preventing coastal erosion and property damage. Rehabilitating mangrove forests can involve planting mangrove saplings, removing debris, and ensuring sustainable land-use practices around these areas. Mangroves also provide essential breeding grounds for fish and other marine life, supporting local fisheries and livelihoods. By investing in mangrove restoration, coastal cities can enhance their natural defence mechanisms while promoting economic opportunities through sustainable fisheries and eco-tourism.

Protecting Coral Reefs

Coral reefs act as underwater barriers that dissipate wave energy, reducing the impact of storms on coastal areas. Protecting and restoring coral reefs can involve measures such as coral gardening, where fragments of healthy corals are grown in nurseries and then transplanted to degraded reef areas. Efforts to reduce pollution, overfishing, and destructive coastal development are also crucial for the health of coral reefs. Healthy reefs support marine biodiversity, fisheries, and tourism, providing economic and ecological benefits. Investing in coral reef protection helps safeguard coastal communities from the adverse effects of sea-level rise and promotes marine conservation.

Implementing Hybrid Solutions

Hybrid solutions combine natural and engineered approaches to enhance coastal resilience. For example, integrating vegetation into the design of constructed levees can provide additional stability and reduce erosion. Living shorelines, which use plants, sand, and rock to stabilize coastlines, offer an alternative to traditional hard infrastructure.

These solutions can adapt to changing environmental conditions and provide habitats for wildlife. Hybrid approaches can also include the construction of oyster reefs that protect shorelines while filtering water and supporting marine ecosystems. By blending natural and artificial elements, hybrid solutions offer flexible and adaptive strategies for coastal protection.

Enhancing Biodiversity and Supporting Fisheries

Natural and hybrid solutions not only protect coastal cities but also enhance biodiversity and support fisheries. Wetlands, mangroves, and coral reefs are rich in biodiversity, providing habitats for a wide range of species. These ecosystems support the food web, including commercially important fish species, thereby sustaining local fisheries and contributing to food security. Healthy ecosystems also attract tourism, creating economic opportunities for local communities. Investing in these natural systems fosters a symbiotic relationship between environmental protection and economic development, ensuring long-term sustainability.

Cost-Effectiveness and Sustainability

Natural and hybrid solutions are often more cost-effective and sustainable compared to traditional engineered structures. While the initial investment in restoration or hybrid infrastructure may be significant, these solutions typically have lower maintenance costs and longer lifespans. Natural systems can self-repair and adapt to environmental changes, reducing the need for frequent and expensive repairs. Additionally, the co-benefits of ecosystem services, such as water filtration, carbon sequestration, and recreational opportunities, provide added value beyond flood protection. These attributes make natural and hybrid solutions attractive options for coastal cities seeking resilient and sustainable adaptation measures.

Community and Stakeholder Engagement

Successful implementation of natural and hybrid solutions require the involvement of local communities and stakeholders. Engaging residents in restoration projects fosters a sense of ownership and stewardship, enhancing the long-term sustainability of these efforts. Educational programmes and public outreach can raise awareness about the benefits of natural and hybrid solutions, encouraging community support and participation. Collaborative efforts with local governments, NGOs, and private sector partners can leverage resources and expertise, ensuring comprehensive and effective adaptation strategies.

Conclusion

In conclusion, investing in natural and hybrid solutions provides coastal cities with a resilient and cost-effective means of mitigating the risks associated with sea-level rise. Restoring wetlands, rehabilitating mangroves, protecting coral reefs, and implementing hybrid approaches offer multifaceted benefits, including enhanced flood protection, biodiversity conservation, and support for local fisheries. These strategies not only shield coastal communities from the immediate impacts of climate change but also contribute to long-term environmental sustainability and economic prosperity. By prioritizing natural and hybrid solutions, coastal cities can build resilience, support ecosystems, and foster community engagement, paving the way for a sustainable and secure future.

Cost Implications and Funding Mechanisms

Implementing strategies to mitigate the impacts of sea-level rise in coastal cities involves substantial financial commitments. These costs can vary widely depending on the specific measures adopted, the scale of the projects, and the geographical and socio-economic context of the area. To ensure the effectiveness and sustainability of these adaptation efforts, a range of funding mechanisms can be leveraged.

- **Cost Implications of Adaptation Measures:** The financial implications of adaptation strategies are significant and multifaceted. Constructing physical defences such as seawalls, levees, and floodgates can require enormous capital investments. For instance, the Thames Barrier in London, designed to protect the city from tidal surges, cost approximately £1.6 billion to construct and requires ongoing maintenance expenditures. Similarly, the Maeslantkering in Rotterdam represents a major engineering feat with substantial construction and operational costs.

These large-scale infrastructure projects, while effective, necessitate long-term financial commitments to ensure their functionality and resilience against escalating climate impacts. In addition to the direct costs of construction and maintenance, there are indirect costs associated with these projects. These can include land acquisition costs, potential displacement of communities, and environmental mitigation expenses. Moreover, the need for continuous monitoring and upgrading of these defences to cope with rising sea levels and increased storm intensity adds to the overall financial burden.

- **Natural and Hybrid Solutions:** While natural and hybrid solutions such as restoring wetlands, mangroves, and coral reefs are generally more cost-effective in the long run, they still require significant upfront investments.

  For example, wetland restoration involves costs related to land acquisition, replanting, and hydrological modifications. However, these natural systems can offer cost savings over time due to their ability to self-repair and adapt, reducing the need for expensive maintenance and repairs typically associated with hard infrastructure. Hybrid solutions, which combine natural elements with engineered structures, also involve substantial costs. These projects, such as living shorelines that integrate vegetation with hard materials like rock and sand, require careful planning and design to balance the ecological benefits with the engineering requirements. The costs of hybrid solutions can be offset by their multifunctional benefits, including habitat creation, water filtration, and recreational opportunities.

- **Implementing Land-use Planning and Zoning Regulations:** Effective land-use planning and zoning regulations are essential for reducing vulnerability to sea-level rise but also involve financial implications. Developing and enforcing these regulations require investments in data collection, mapping, and modelling to identify high-risk areas. Additionally, local governments may need to provide incentives for relocation or redevelopment in safer areas, which can entail compensation costs for property owners and businesses. Public outreach and education campaigns are also necessary to ensure community buy-in and compliance with new regulations.

- **Emergency Response and Recovery Costs:** The increasing frequency and intensity of coastal flooding and storms necessitate substantial investments in emergency response and recovery efforts. These include costs for disaster preparedness, such as developing evacuation plans, conducting drills, and establishing emergency shelters. Post-disaster recovery costs can be enormous, involving rebuilding infrastructure, restoring services, and providing support to affected communities. For instance, the recovery efforts after Hurricane Sandy in New York City were estimated to cost over $50 billion, highlighting the immense financial burden of such events.

Funding Mechanisms

To manage the substantial costs associated with adapting to sea-level rise, coastal cities can utilize a variety of funding mechanisms. These include government funding, public-private partnerships, insurance and risk transfer mechanisms, and innovative financing options such as climate bonds.

- **Government Funding:** Government funding is a primary source of financing for adaptation projects. Local, state, and national governments can allocate budgetary resources for constructing and maintaining physical defences, implementing natural and hybrid solutions, and developing land-use planning regulations. Governments can also apply for grants and funding from international organizations and climate funds, such as the Green Climate Fund (GCF). These funds provide financial assistance to developing countries for climate adaptation and mitigation projects, helping to offset the high costs of implementation.

- **Public-Private Partnerships (PPPs):** PPPs are effective in leveraging private sector investment and expertise to fund and implement adaptation projects. These partnerships can pool resources from both the public and private sectors, sharing risks and benefits. For example, private companies can invest in the construction and maintenance of seawalls or floodgates in exchange for concessions or revenue-sharing agreements. PPPs can also facilitate innovation and efficiency in project delivery, ensuring that adaptation measures are implemented cost-effectively.
- **Insurance and Risk Transfer Mechanisms:** Innovative insurance products and risk transfer mechanisms can provide financial protection against the impacts of sea-level rise. Parametric insurance, for example, pays out based on predefined triggers such as specific weather events, providing quick financial relief for recovery and rebuilding efforts. Catastrophe bonds (cat bonds) allow governments to transfer the risk of severe climate events to investors, who receive payouts if the events do not occur, thus securing funds for post-disaster reconstruction. These mechanisms can help manage the financial risks associated with climate impacts and ensure that funds are available when needed.
- **Climate Bonds and Green Financing:** Issuing climate bonds and securing green financing can raise capital for large-scale adaptation projects. Climate bonds are financial instruments designed to fund projects that have positive environmental benefits, including those aimed at mitigating the impacts of sea-level rise. These bonds attract investors interested in supporting sustainable and resilient infrastructure development. Green financing initiatives, supported by international organizations and financial institutions, provide funding for projects that promote environmental sustainability and resilience.

- **Innovative Local Financing:** Local governments can explore innovative financing mechanisms tailored to their specific needs and contexts. For example, special assessment districts can be established to fund adaptation projects through property taxes or fees levied on property owners in vulnerable areas. Tax increment financing (TIF) allows cities to use future gains in property taxes resulting from increased property values to finance current adaptation projects. Local governments can also create resilience funds, pooling resources from various stakeholders to support long-term adaptation efforts.
- **International Aid and Development Assistance:** Developing countries facing significant adaptation challenges can benefit from international aid and development assistance. International organizations, such as the World Bank and the United Nations, provide grants, loans, and technical assistance to support adaptation projects. These funds can help offset the costs of implementing adaptation measures, ensuring that vulnerable communities have the resources needed to enhance their resilience to sea-level rise.
- **Philanthropic Contributions and Community Funding:** Philanthropic organizations and community-based funding initiatives can also play a role in financing adaptation projects. Foundations and non-profit organizations can provide grants and donations to support local adaptation efforts, particularly those that benefit marginalized and vulnerable populations. Community funding mechanisms, such as crowdfunding and local investment funds, can mobilize resources from residents and local businesses to finance small-scale adaptation projects.

Conclusion

The cost implications of adapting to sea-level rise in coastal cities are significant, requiring substantial investments in physical defences, natural and hybrid solutions, land-use planning, and emergency response and recovery efforts. To manage these costs and ensure the successful implementation of adaptation measures, a variety of funding mechanisms can be utilized. Government funding, public-private partnerships, insurance and risk transfer mechanisms, climate bonds, and green financing are essential tools for mobilizing the necessary resources. By leveraging these funding mechanisms and exploring innovative local and international financing options, coastal cities can build resilience, protect their populations and economies, and ensure a sustainable and secure future in the face of rising sea levels.

Case Studies of Coastal Adaptation Projects

Rotterdam, Netherlands

Rotterdam is a leading example of innovative coastal adaptation. The city has implemented a comprehensive adaptation strategy that includes the construction of advanced flood defences, such as the Maeslantkering storm surge barrier, and the development of "water plazas" that temporarily store rainwater during heavy rainfall. Rotterdam's approach combines engineering solutions with sustainable urban planning, creating a resilient and livable city.

- **Maeslantkering Storm Surge Barrier:** The Maeslantkering is one of the largest movable flood barriers in the world. It protects the port of Rotterdam, one of the busiest ports globally, and the surrounding areas from storm surges. The barrier consists of two massive gates that can be closed during extreme weather conditions to prevent seawater from flooding the city. The construction of the Maeslantkering, completed in 1997, was a significant engineering feat and part of the broader Delta Works project, a series of dams, sluices, locks, dikes, and storm surge barriers designed to protect the Netherlands from the sea. This infrastructure provides robust protection against rising sea levels and storm surges, ensuring the safety and economic stability of Rotterdam.

- **Water Plazas:** In addition to large-scale engineering solutions, Rotterdam has implemented innovative urban design strategies, such as the development of water plazas. These plazas serve a dual purpose: they act as public spaces during dry weather and as temporary water storage areas during heavy rainfall. Water plazas help to manage urban runoff, reducing the risk of flooding in densely populated areas. They are designed to collect excess rainwater, which is then gradually released into the city's drainage system. This approach not only mitigates flood risks but also enhances the quality of urban life by providing green, multifunctional spaces for residents.
- **Urban Planning and Resilience:** Rotterdam's adaptation strategy extends beyond physical infrastructure to include comprehensive urban planning. The city has integrated resilience into its spatial planning processes, ensuring that new developments are designed to withstand climate impacts. This includes elevating buildings, designing flood-proof structures, and creating green roofs and permeable surfaces to enhance water absorption. Rotterdam's commitment to resilient urban planning is evident in its Climate Adaptation Strategy, which outlines measures to reduce vulnerability and increase the city's ability to cope with climate change.
- **Community Engagement and Awareness:** Engaging the local community has been a crucial component of Rotterdam's adaptation efforts. The city has involved residents, businesses, and stakeholders in the planning and implementation of adaptation measures. Public awareness campaigns and educational programmes have been conducted to inform the community about the risks of climate change and the importance of adaptation. By fostering a culture of resilience and involving the

community in decision-making processes, Rotterdam has ensured broad support for its adaptation initiatives.

- **Innovative Financing and Partnerships Financing:** Rotterdam's ambitious adaptation projects have required innovative approaches and strong partnerships. The city has leveraged funding from national and European sources, including the Dutch government and the European Union. Public-Private Partnerships have also played a critical role, bringing together resources and expertise from various sectors. For example, private companies have been involved in the development and maintenance of water plazas and other green infrastructure projects. This collaborative approach has enabled Rotterdam to secure the necessary funding and technical support for its adaptation strategy.
- **Sustainable Development Goals (SDGs):** Rotterdam's adaptation efforts align with several United Nations Sustainable Development Goals (SDGs), including SDG 11 (Sustainable Cities and Communities) and SDG 13 (Climate Action). By investing in resilient infrastructure and sustainable urban planning, Rotterdam is contributing to global efforts to build resilient cities and mitigate the impacts of climate change. The city's innovative solutions serve as a model for other urban centres facing similar challenges, demonstrating how integrated and forward-thinking approaches can enhance urban resilience.
- **Monitoring and Evaluation:** Rotterdam has established robust monitoring and evaluation mechanisms to assess the effectiveness of its adaptation measures. The city regularly reviews and updates its Climate Adaptation Strategy based on new data and climate projections. This iterative approach ensures that the city's adaptation efforts remain relevant and effective in the face of evolving climate risks. By continuously monitoring the performance of its flood defences and urban infrastructure, Rotterdam can make informed decisions and adjustments to enhance its resilience.

- **Challenges and Future Directions:** Despite its successes, Rotterdam faces ongoing challenges in adapting to climate change. The city must continue to invest in the maintenance and upgrading of its flood defences and urban infrastructure to cope with rising sea levels and more frequent extreme weather events. Additionally, Rotterdam must address social equity concerns, ensuring that all communities benefit from adaptation measures. The city is exploring new technologies and approaches, such as smart flood management systems and nature-based solutions, to enhance its resilience further. By remaining proactive and adaptive, Rotterdam can continue to protect its residents and economy from climate impacts.

Conclusion

Rotterdam's comprehensive and innovative approach to coastal adaptation demonstrates the effectiveness of combining engineering solutions with sustainable urban planning and community engagement. The city's implementation of advanced flood defences, such as the Maeslantkering storm surge barrier, and multifunctional urban designs like water plazas, has significantly enhanced its resilience to rising sea levels and extreme weather events. Rotterdam's experience highlights the importance of long-term planning, innovative financing, and stakeholder involvement in building resilient and livable coastal cities. As climate change continues to pose significant risks, other coastal cities can learn from Rotterdam's example to develop integrated and forward-thinking adaptation strategies that protect their communities and support sustainable development.

New York City, USA: Coastal Adaptation and Resilience

- **New York City, USA:** After Hurricane Sandy in 2012, New York City launched several initiatives to enhance its resilience to sea-level rise and extreme weather events. The city's "Rebuild by Design" competition generated innovative solutions, such as the "Big U" project, which proposes a protective system of berms and parks around Lower Manhattan. New York has also invested in upgrading its infrastructure, improving building codes, and restoring natural buffers like wetlands.

- **Rebuild by Design Competition:** In response to the devastating impacts of Hurricane Sandy, New York City initiated the "Rebuild by Design" competition in 2013. This initiative aimed to develop innovative and sustainable strategies to protect the city from future climate-related disasters. The competition invited architects, designers, engineers, and planners from around the world to propose creative solutions for enhancing the city's resilience. One of the most prominent projects to emerge from this competition is the "Big U" project.
- **The Big U Project:** The "Big U" is a comprehensive flood protection system designed to shield Lower Manhattan from storm surges and sea-level rise. This project envisions a series of interconnected berms, parks, and public spaces that act as both recreational areas and protective barriers. The design includes features such as elevated landscapes, floodwalls, and water-retaining areas that can absorb and divert floodwaters. The "Big U" not only enhances flood protection but also improves the urban environment by providing green spaces and recreational facilities for residents and visitors. This innovative approach integrates infrastructure with public amenities, promoting both safety and quality of life.
- **Infrastructure Upgrades:** In addition to the "Big U", New York City has made significant investments in upgrading its infrastructure to withstand extreme weather events. The city has focused on strengthening its transportation networks, utilities, and public buildings. For example, subway stations and tunnels have been retrofitted with flood protection measures, such as watertight doors and

raised ventilation systems, to prevent inundation during storms. Critical infrastructure, including power plants and wastewater treatment facilities, has been elevated or relocated to reduce vulnerability to flooding.

- **Improving Building Codes:** To ensure new developments are resilient to climate impacts, New York City has revised its building codes and zoning regulations. The updated codes require that new buildings in flood-prone areas be elevated above predicted flood levels and incorporate flood-resistant materials and designs. These regulations also mandate that essential services, such as electrical systems and emergency generators, be located above potential flood levels to maintain functionality during extreme weather events. By enforcing stricter building standards, the city aims to reduce the risk of damage and enhance the safety of its residents.
- **Restoring Natural Buffers:** Recognizing the importance of natural ecosystems in flood mitigation, New York City has invested in restoring and preserving natural buffers like wetlands and coastal marshes. These natural areas act as sponges, absorbing floodwaters and reducing the impact of storm surges on built environments. Projects such as the restoration of Jamaica Bay aim to enhance the ecological health of these areas while providing critical flood protection. The city's efforts include replanting native vegetation, removing invasive species, and improving water quality through better management practices. Restoring natural buffers not only protects against flooding but also supports biodiversity and recreational opportunities.
- **Community Engagement and Preparedness:** New York City has prioritized community engagement and preparedness as key components of its resilience strategy. The city has conducted extensive outreach to educate residents about the risks of sea-level rise and extreme weather events. Public awareness campaigns and

community workshops have been organized to inform residents about evacuation routes, emergency procedures, and ways to protect their homes and businesses. The city has also developed community-based resilience plans, involving local stakeholders in the planning and implementation of adaptation measures. By fostering a culture of preparedness and resilience, New York City aims to empower its residents to take proactive steps in mitigating climate risks.

- **Financial Mechanisms and Partnerships:** Funding the extensive adaptation projects in New York City has required a combination of public and private financing. The city has secured federal grants, state funding, and municipal bonds to support its resilience initiatives. Public-Private Partnerships have also been instrumental in mobilizing resources and expertise for large-scale projects. For example, collaborations with private developers have facilitated the integration of flood protection measures into new developments, ensuring that they contribute to the city's overall resilience. Innovative financing mechanisms, such as resilience bonds, have been explored to attract investment for climate adaptation projects.
- **Monitoring and Adaptive Management:** To ensure the effectiveness of its adaptation efforts, New York City has implemented robust monitoring and adaptive management practices. The city regularly assesses the performance of its infrastructure and resilience measures, using data and feedback to make necessary adjustments. This iterative approach allows the city to respond to new information and changing conditions, ensuring that its adaptation strategies remain effective over time. Monitoring programmes include tracking sea-level rise, storm surge patterns, and the condition of natural buffers, providing critical data for informed decision-making.

- **Challenges and Future Directions:** Despite significant progress, New York City faces ongoing challenges in its adaptation efforts. The scale and complexity of the city's infrastructure require continuous investment and maintenance to keep pace with rising sea levels and more frequent extreme weather events. Social equity concerns must also be addressed, ensuring that all communities, particularly vulnerable and low-income populations, benefit from resilience initiatives. The city is exploring new technologies and approaches, such as green infrastructure and advanced modelling techniques, to enhance its resilience further. By remaining proactive and adaptive, New York City aims to continue protecting its residents and economy from the impacts of climate change.

Conclusion

New York City's comprehensive and multifaceted approach to coastal adaptation demonstrates the importance of innovative design, infrastructure upgrades, regulatory improvements, and community engagement in building resilience to sea-level rise and extreme weather events. Initiatives such as the "Rebuild by Design" competition and the "Big U" project highlight the potential for creative solutions that integrate flood protection with urban development. By investing in infrastructure, improving building codes, restoring natural buffers, and fostering community preparedness, New York City is enhancing its ability to withstand and recover from climate impacts. The city's experience serves as a valuable model for other coastal cities seeking to develop robust and inclusive adaptation strategies that ensure long-term sustainability and safety.

Venice, Italy

Venice faces unique challenges due to its location and historical significance. The MOSE (Modulo Sperimentale Elettromeccanico) project involves a system of mobile barriers designed to protect the city from high tides and flooding. When not in use, the barriers lie on the seabed, but they can be raised to block incoming tides. The project aims to preserve Venice's cultural heritage while protecting its residents and economy.

- **MOSE Project:** The MOSE project, one of the most ambitious flood defence systems in the world, aims to safeguard Venice from the increasingly frequent and severe high tides known as acqua alta. The system consists of 78 mobile barriers installed at the three inlets of the Venetian Lagoon—Lido, Malamocco, and Chioggia.
  These barriers are normally submerged on the seabed but can be inflated with air to rise above the water surface, effectively blocking incoming tides and preventing flooding in Venice.
- **Engineering and Design:** The design of the MOSE project reflects a combination of advanced engineering and sensitivity to the unique environmental and cultural context of Venice. The barriers are designed to operate under various tidal conditions, ensuring that they can protect the city even during severe storms. The system is also designed to minimize environmental impact, with the barriers remaining hidden underwater when not in use, preserving the aesthetic and ecological integrity of the lagoon.
- **Cultural and Economic Preservation:** Venice's economy heavily relies on tourism, and its historical and architectural heritage is a major draw for visitors. Frequent flooding not only damages the city's infrastructure and buildings but also disrupts tourism and the local economy. The MOSE project aims to provide a long-term solution to these challenges, preserving Venice's cultural heritage and ensuring that the city remains an attractive destination for tourists. By protecting its iconic landmarks and historic sites from flood damage, the project supports the sustainability of Venice's tourism industry.

- **Challenges and Criticisms:** Despite its ambitious goals, the MOSE project has faced several challenges and criticisms. The project has experienced significant delays and cost overruns, with its completion postponed multiple times. Additionally, concerns have been raised about the potential environmental impacts of the barriers, including changes to the lagoon's natural water exchange and sediment dynamics. There have also been issues related to maintenance and operational costs, which require ongoing funding and technical expertise to ensure the system's effectiveness.
- **Future Directions:** As the MOSE project nears full operational status, the focus will shift to optimizing its performance and addressing any emerging issues. This includes regular maintenance, monitoring environmental impacts, and ensuring that the system can adapt to future sea-level rise projections. The project represents a critical component of Venice's broader strategy to adapt to climate change and protect its cultural and economic assets.

Jakarta, Indonesia

Jakarta is addressing its severe flooding problems with a combination of measures, including the construction of seawalls, river dredging, and the relocation of vulnerable communities.

The city's "Giant Sea Wall" project aims to protect Jakarta from rising sea levels while creating new urban spaces through land reclamation. This ambitious plan seeks to balance economic development with environmental sustainability.

- **The Giant Sea Wall Project:** Jakarta's "Giant Sea Wall" project, officially known as the National Capital Integrated Coastal Development (NCICD) project, is a multi-phase initiative designed to protect the city from chronic flooding caused by sea-level rise, land subsidence, and heavy rainfall. The project involves constructing an extensive seawall along Jakarta's coastline, complemented by a series of smaller barriers and water management systems within the city.

- **Land Reclamation and Urban Development:** In addition to providing flood protection, the Giant Sea Wall project includes significant land reclamation efforts to create new urban spaces. These reclaimed areas are intended to support economic development by providing land for housing, commercial activities, and infrastructure. The project aims to transform Jakarta's coastline into a vibrant and resilient urban area, capable of withstanding environmental challenges while supporting the city's growing population and economy.
- **Relocation of Vulnerable Communities:** An essential component of Jakarta's adaptation strategy is the relocation of communities living in high-risk areas. Many residents in low-lying and flood-prone neighbourhoods face constant threats from rising waters and poor living conditions. The city government has implemented programmes to relocate these communities to safer areas, providing them with improved housing and access to essential services. This approach not only enhances the safety and well-being of vulnerable populations but also reduces the strain on emergency response systems during flood events.
- **River Dredging and Water Management:** To address flooding from heavy rainfall and river overflow, Jakarta has undertaken extensive river dredging and water management initiatives. These efforts aim to improve the capacity and flow of the city's waterways, reducing the risk of urban flooding. The dredging projects involve removing sediment and debris from rivers and canals, enhancing their ability to convey water efficiently. Additionally, the city has invested in upgrading drainage systems and constructing retention basins to manage storm water more effectively.
- **Environmental and Social Considerations:** The Giant Sea Wall project and associated adaptation measures must

balance economic development with environmental and social considerations. Land reclamation and infrastructure projects can have significant ecological impacts, including habitat loss and changes to coastal ecosystems. To mitigate these effects, the project includes environmental management plans and measures to restore and conserve natural habitats. Social considerations, such as ensuring fair compensation and support for relocated communities, are also crucial to the project's success.

- **Challenges and Implementation:** Jakarta's adaptation efforts face several challenges, including securing adequate funding, coordinating among multiple stakeholders, and addressing the technical complexities of large-scale infrastructure projects. The Giant Sea Wall project, in particular, requires significant investment from both public and private sources. Coordination among government agencies, international partners, and local communities is essential to ensure the project's smooth implementation and long-term sustainability.
- **Future Directions:** As Jakarta continues to develop and implement its adaptation strategies, ongoing monitoring and evaluation will be critical to success. The city must remain flexible and adaptive, responding to new data and emerging challenges related to climate change. Strengthening community engagement and building local capacity for disaster risk reduction are also key priorities. By integrating resilience into its urban planning and development processes, Jakarta aims to protect its residents and economic assets from the impacts of sea-level rise and flooding.

Conclusion

The coastal adaptation projects in Venice and Jakarta highlight the diverse strategies cities can adopt to address the challenges posed by sea-level rise and flooding. Venice's MOSE project exemplifies the use of advanced engineering to protect a city with significant cultural heritage, while Jakarta's comprehensive approach integrates large-scale infrastructure, urban development, and community relocation to enhance resilience. Both cities face ongoing challenges in financing, implementation, and balancing environmental and social impacts, but their efforts provide valuable lessons for other coastal cities seeking to develop sustainable adaptation strategies.

By investing in innovative and multifaceted solutions, cities can safeguard their populations, economies, and cultural assets from the adverse effects of climate change. Coastal cities face significant economic risks from rising sea levels, but they can employ various strategies to mitigate these risks. Building physical defences, implementing land-use planning and zoning regulations, and investing in natural and hybrid solutions are crucial for protecting coastal infrastructure, properties, and economies. While the cost implications of these measures are substantial, diverse funding mechanisms, including government funding, PPPs, insurance products, and climate bonds, can support adaptation efforts. Case studies from Rotterdam, New York City, Venice, and Jakarta demonstrate that innovative and proactive approaches can enhance the resilience of coastal cities, safeguarding them against the impacts of climate change. By learning from these examples and investing in comprehensive adaptation strategies, coastal cities can mitigate the risks of rising seas and secure a sustainable future.

REFERENCES

Daly, M.C. and Hobijn, B. (2022). "Wage Growth and Inflation: The Role of Productivity and Labour Market Slack". *Economic Review, Federal Reserve Bank of Kansas City, 2022*(Q2): 47-74.

Draghi, M. (2022). "Policy Coordination in a High-Inflation Environment". *Journal of International Economics*, Vol. 132, Editorial Board Article No. 103594.

Drechsler, I., Savov, A. and Schnabl, P. (2022). "The Financial Stability Implications of Inflation". *Journal of Finance, 77*(4): 1873-1911.

Eggertsson, G.B. and Woodford, M. (2022). "The Zero Bound on Interest Rates and Optimal Monetary Policy". *Brookings Papers on Economic Activity, 2022*(2): 139-211.

Guvenen, F., Kaplan, G. and Song, J. (2022). "How Does Inflation Affect Different Income Groups?" *Journal of Political Economy, 130*(5): 1437-1473.

Hamilton, J.D. (2022). "Why You Should Never Use the Hodrick-Prescott Filter". *Review of Economic Studies, 89*(1): 104-120.

Jordà, Ò. and Taylor, A.M. (2022). "The Pandemic, Inflation, and Debt Sustainability". *Economic Policy, 37*(110): 275-314.

Kiley, M.T. (2022). "Unemployment, Inflation, and Monetary Policy: Evidence from Advanced and Emerging Market Economies". *Journal of International Money and Finance*, Vol. 122, Editorial Board Article No. 102564.

Laubach, T. and Williams, J.C. (2022). "Measuring the Natural Rate of Interest: International Trends and Determinants". *Brookings Papers on Economic Activity, 2022*(3): 57-108.

Martin, F. (2022). "Inflation, Wage Setting, and the Monetary Transmission Mechanism". *Journal of Economic Theory*, Vol. 197, Editorial Board Article No. 105299.

Mian, A., Sufi, A. and Verner, E. (2022). "Household Debt and Inflation: The Role of Mortgage Debt in Monetary Policy Transmission". *Journal of Financial Economics, 143*(3): 1041-1061.

Mishkin, F.S. (2022). "Monetary Policy Strategy: Lessons from the Financial Crisis". *Journal of Economic Perspectives, 36*(3): 123-148.

Mussa, M. (2022). "Inflation, Real Interest Rates, and the Real Exchange Rate: An Empirical Investigation". *Journal of Monetary Economics, 127*: 243-271.

Orphanides, A. (2022). "Policy Rules and Economic Stability: Lessons from the Great Inflation". *Journal of Economic Dynamics and Control*, Vol. 135, Editorial Board Article No. 104313.

Reis, R. (2022). "The Analytics of Monetary Policy: A New Keynesian Perspective". *Quarterly Journal of Economics*, 137(4): 1951-1987.

Shiller, R.J. (2022). *Narrative Economics: How Stories Go Viral and Drive Major Economic Events*, 2nd edn., Princeton University Press.

Sims, C.A. (2022). "Fiscal Policy, Inflation, and Debt Sustainability". *Journal of Economic Theory*, Vol. 198, Editorial Board Article No. 105297.

Smets, F. and Wouters, R. (2022). "An Estimated Dynamic Stochastic General Equilibrium Model of the Euro Area". *Journal of Economic Dynamics and Control*, Vol. 133, Editorial Board Article No. 104295.

Stock, J.H. and Watson, M.W. (2022). "Disentangling the Channels of the 2021-2022 Inflation Surge". *Journal of Applied Econometrics, 37*(6): 975-996.

Summers, L.H. (2022). "Secular Stagnation and the Road to Recovery". *American Economic Review: Papers & Proceedings, 112*: 104-109.

Woodford, M. (2022). "Inflation and Financial Stability: A Microeconomic Perspective". *Journal of Economic Literature, 60*(2): 337-365.

9

Economic Impacts of Extreme Weather Events

Lessons Learned and Future Preparedness

Analysis of Past Extreme Weather Events

Extreme weather events, such as hurricanes, floods, droughts, and wildfires, have become increasingly frequent and severe due to climate change. Analyzing past events provides valuable insights into their economic impacts and informs future preparedness strategies.

- **Hurricane Katrina (2005):** Hurricane Katrina is one of the most devastating hurricanes in U.S. history, causing widespread destruction in New Orleans and the Gulf Coast. The storm resulted in over 1,800 deaths and caused approximately $125 billion in damages. The economic impact was profound, affecting various sectors, including housing, infrastructure, and tourism. The failure of levees and inadequate emergency response highlighted the need for robust infrastructure and effective disaster management systems.

- **Japanese Tsunami (2011):** The earthquake and tsunami that struck Japan in 2011 caused significant economic disruption, particularly in the manufacturing and energy sectors. The disaster led to the shutdown of the Fukushima Daiichi nuclear plant, resulting in long-term energy shortages and a shift in energy policy. The total economic cost was estimated at $235 billion, making it one of the costliest natural disasters in history. The event underscored the importance of building resilient infrastructure and diversifying energy sources to mitigate future risks.
- **European Heat Wave (2003):** The 2003 European heat wave resulted in over 70,000 deaths and had severe economic impacts, particularly on agriculture and public health. Crops failed across large parts of Europe, leading to significant losses in the agricultural sector. The heat wave also strained healthcare systems, with increased hospital admissions and heat-related illnesses. This event highlighted the need for improved heat wave preparedness, including early warning systems and infrastructure designed to cope with extreme temperatures.
- **Australian Bushfires (2019-20):** The Australian bushfires of 2019-20, known as Black Summer, caused widespread devastation, burning over 18 million hectares, destroying thousands of homes, and killing at least 33 people. The economic impact was estimated at around $103 billion, affecting agriculture, tourism, and air quality.
  The fires highlighted the increasing risk of extreme weather events exacerbated by climate change and underscored the need for better forest management, fire fighting resources, and community preparedness.
- **Pakistan Floods (2010):** The 2010 floods in Pakistan were among the worst in the country's history, affecting over 20 million people and causing economic losses estimated at $43 billion. The floods damaged infrastructure, agriculture, and homes, leading to long-term economic and social

challenges. The disaster emphasized the importance of effective water management, resilient infrastructure, and robust emergency response systems.

Economic Impacts on Different Sectors

Extreme weather events affect various economic sectors differently, with some sectors experiencing more immediate and severe impacts than others.

- **Agriculture:** Agriculture is particularly vulnerable to extreme weather events. Droughts can lead to crop failures and water shortages, while floods can destroy crops and erode fertile soil. For instance, the 2012 drought in the United States caused significant reductions in corn and soybean yields, leading to higher food prices and financial losses for farmers. The agricultural sector must invest in resilient farming practices, such as drought-resistant crops and efficient irrigation systems, to mitigate these risks.
- **Infrastructure:** Extreme weather can cause widespread damage to infrastructure, including roads, bridges, and buildings. For example, Hurricane Sandy in 2012 caused extensive damage to New York City's infrastructure, resulting in billions of dollars in repair and recovery costs. Ensuring infrastructure is resilient to extreme weather involves investing in durable materials, elevated structures, and improved drainage systems.
- **Tourism:** The tourism industry is highly sensitive to weather conditions. Natural disasters can deter tourists, leading to significant revenue losses. Following the 2004 Indian Ocean tsunami, many coastal resorts in affected countries experienced a sharp decline in visitors, impacting local economies dependent on tourism. Diversifying tourism offerings and enhancing disaster preparedness can help mitigate these impacts.
- **Energy:** Extreme weather events can disrupt energy production and distribution. The 2011 Japanese tsunami,

for example, led to the shutdown of several nuclear reactors, resulting in energy shortages and increased reliance on fossil fuels. Building a resilient energy infrastructure that includes renewable energy sources and robust distribution networks is crucial for maintaining energy security during extreme weather.

- **Insurance:** The insurance sector faces increased claims and financial losses following extreme weather events. Insurers must adjust their risk models and pricing strategies to account for the growing frequency and severity of these events. For example, the increased frequency of hurricanes has led to higher premiums and stricter underwriting standards for properties in vulnerable areas. Developing innovative insurance products, such as parametric insurance, can provide more efficient risk transfer mechanisms.

Preparedness and Response Strategies

Effective preparedness and response strategies are essential for minimizing the economic impacts of extreme weather events.

- **Early Warning Systems:** Implementing early warning systems that provide timely and accurate information about impending extreme weather events can save lives and reduce economic losses. For example, advanced meteorological systems and communication networks can alert communities to evacuate or take protective measures before a hurricane or flood.
- **Disaster Risk Reduction (DRR) Planning:** Comprehensive DRR planning involves identifying vulnerabilities, assessing risks, and implementing measures to mitigate them.
  This includes building resilient infrastructure, developing evacuation plans, and conducting regular disaster preparedness drills. Community engagement and

education are also critical components of effective DRR planning.

- **Climate-Resilient Infrastructure:** Investing in climate-resilient infrastructure is crucial for reducing the economic impacts of extreme weather. This includes constructing buildings and roads that can withstand floods, storms, and heat waves. For example, incorporating green infrastructure, such as green roofs and permeable pavements, can enhance urban resilience by managing storm water and reducing heat island effects.
- **Emergency Response and Recovery:** Efficient emergency response and recovery systems are essential for minimizing the immediate and long-term economic impacts of extreme weather. This includes ensuring that emergency services are well-equipped and trained to respond to disasters, as well as establishing recovery funds to support rebuilding efforts. For instance, the Federal Emergency Management Agency (FEMA) in the United States plays a crucial role in coordinating disaster response and providing financial assistance for recovery.

Future Projections and Economic Planning

Looking ahead, it is crucial to integrate future climate projections into economic planning to enhance preparedness and resilience.

- **Climate Modelling and Projections:** Utilizing advanced climate models to project future weather patterns and extreme events can inform long-term economic planning. These models can help identify areas at greatest risk and guide investments in infrastructure and adaptation measures. For example, cities like New York and Rotterdam use climate projections to plan for future sea-level rise and increased storm intensity.
- **Sustainable Development Goals (SDGs):** Aligning economic planning with the United Nations Sustainable

Development Goals (SDGs) can promote resilience and sustainability. Goals such as SDG 9 (Industry, Innovation, and Infrastructure) and SDG 13 (Climate Action) emphasize the importance of building resilient infrastructure and taking urgent action to combat climate change.

- **Public-Private Partnerships (PPPs):** Collaborating with the private sector through PPPs can mobilize resources and expertise for climate adaptation and resilience projects. These partnerships can facilitate investments in resilient infrastructure, renewable energy, and innovative technologies. For instance, the Rockefeller Foundation's 100 Resilient Cities initiative has partnered with cities worldwide to develop and implement resilience strategies.
- **Insurance and Financial Instruments:** Developing innovative insurance products and financial instruments can provide financial protection against the economic impacts of extreme weather. This includes expanding the use of climate bonds, catastrophe bonds, and parametric insurance to transfer risks and fund recovery efforts. For example, the Caribbean Catastrophe Risk Insurance Facility (CCRIF) provides parametric insurance coverage to Caribbean governments, enabling quick payouts after extreme weather events.
- **Capacity Building and Education:** Building the capacity of communities, businesses, and governments to prepare for and respond to extreme weather is essential for long-term resilience. This includes training programmes, public awareness campaigns, and the integration of climate risk management into education systems. For instance, initiatives like the United Nations Office for Disaster Risk Reduction (UNDRR) provide training and resources to enhance disaster preparedness globally.

Conclusion

The economic impacts of extreme weather events are profound and far-reaching, affecting various sectors and posing significant challenges for communities and governments. By analyzing past events, understanding sector-specific impacts, and implementing robust preparedness and response strategies, societies can enhance their resilience to these growing threats. Future economic planning must integrate climate projections, align with sustainable development goals, and leverage innovative financing mechanisms to ensure preparedness and resilience. Through collaborative efforts and proactive measures, it is possible to mitigate the economic impacts of extreme weather and build a more resilient and sustainable future.

Economic Impacts on Different Sectors

Extreme weather events affect various economic sectors differently, with some sectors experiencing more immediate and severe impacts than others.

Agriculture: Vulnerability to Extreme Weather Events and Strategies for Resilience

Agriculture, a cornerstone of global economies and a vital source of food security, is particularly vulnerable to the impacts of extreme weather events. These events, intensified by climate change, include droughts, floods, hurricanes, and extreme temperatures. They can cause significant disruptions in agricultural production, leading to economic losses and food shortages. To mitigate these risks, the agricultural sector must adopt resilient farming practices and innovative technologies.

- **Droughts and their Impact on Agriculture:** Droughts are one of the most severe climatic threats to agriculture. They lead to water shortages, reducing the availability of water for irrigation and livestock. This results in crop failures and decreased agricultural productivity. For instance, the 2012 drought in the United States was one of the most severe in recent history. It caused significant reductions in the yields of major crops like corn and soybeans, leading to higher food prices and substantial financial losses for farmers. The drought affected about 80 per cent of U.S. agricultural land, with the total economic impact estimated at $30 billion. The severity of droughts is exacerbated by changing climate patterns, making them more frequent and intense. This trend threatens food security globally, particularly in regions already prone to water scarcity. Developing and implementing strategies to cope with droughts is essential for sustaining agricultural productivity and ensuring food security.

- **Floods and their Devastating Effects:** Floods are another major threat to agriculture, causing widespread destruction of crops and soil erosion. They can lead to the loss of fertile topsoil, reducing the land's productivity for future crop cycles. Floods also damage infrastructure, such as irrigation systems and storage facilities, further complicating recovery efforts.

  For example, the 2010 floods in Pakistan were among the worst in the country's history, affecting over 20 million people and causing economic losses estimated at $43 billion. The floods inundated millions of hectares of agricultural land, destroying crops and displacing farming communities. Recovery from such events is slow and costly, emphasizing the need for effective flood management and mitigation strategies.
- **Hurricanes and Extreme Weather:** Hurricanes and extreme weather events bring a combination of destructive forces, including high winds, heavy rainfall, and storm surges. These events can cause significant damage to agricultural infrastructure, crops, and livestock. For instance, Hurricane Maria in 2017 devastated Puerto Rico's

agricultural sector, causing nearly $780 million in damages. The storm destroyed crops, livestock, and infrastructure, leading to long-term disruptions in food production and supply chains.

The increasing frequency and intensity of such extreme weather events highlight the urgent need for the agricultural sector to adopt more resilient practices and infrastructure to withstand these impacts.

Strategies for Resilient Agriculture

To mitigate the risks associated with extreme weather events, the agricultural sector must invest in resilient farming practices and innovative technologies. Key strategies include:

- **Drought-Resistant Crops:** Developing and planting drought-resistant crop varieties can help sustain agricultural productivity during water shortages. These crops are genetically engineered or selectively bred to withstand prolonged periods of low water availability. For example, drought-resistant maize varieties developed by the International Maize and Wheat Improvement Centre (CIMMYT) have shown promise in maintaining yields under drought conditions.
- **Efficient Irrigation Systems:** Implementing efficient irrigation systems, such as drip irrigation and sprinkler systems, can optimize water use and reduce waste. These systems deliver water directly to the plant roots, minimizing evaporation and runoff. In regions prone to drought, efficient irrigation can significantly enhance water use efficiency and maintain crop yields.
- **Soil Management Practices:** Adopting soil management practices that enhance soil moisture retention and fertility is crucial for resilience against extreme weather. Techniques such as conservation tillage, cover cropping, and organic amendments can improve soil structure and reduce erosion.

These practices also enhance the soil's ability to retain water, making crops more resilient to drought and heavy rainfall.

- **Flood Management and Infrastructure:** Building and maintaining robust flood management infrastructure, such as levees, drainage systems, and floodplains, can protect agricultural land from flooding. Additionally, implementing land-use planning that restricts agricultural activities in flood-prone areas can reduce the risk of crop loss and soil erosion.
- **Climate-Smart Agriculture:** Climate-smart agriculture (CSA) practices integrate sustainable agriculture with climate change adaptation and mitigation. CSA practices include agro-forestry, integrated pest management, and the use of renewable energy in farming operations. These practices enhance the resilience of farming systems to extreme weather while reducing greenhouse gas emissions.
- **Early Warning Systems and Risk Assessment:** Developing and implementing early warning systems for extreme weather events can provide farmers with timely information to take preventive measures. Risk assessment tools that evaluate the vulnerability of agricultural systems to climate impacts can guide decision-making and resource allocation for resilience-building activities.
- **Insurance and Financial Instruments:** Agricultural insurance schemes and financial instruments, such as climate risk insurance and catastrophe bonds, can provide financial protection against the economic losses caused by extreme weather events. These mechanisms help farmers recover more quickly from disasters and invest in resilient practices.

Case Studies and Successful Implementations

- **India's Drought-Resistant Millet Programme:** In India, millet, a drought-resistant grain, has been promoted as a climate-resilient crop. The government has implemented programmes to support millet cultivation through subsidies, research, and market development. As a result, millet production has increased, providing a stable food source and income for farmers in drought-prone regions.

- **Israel's Advanced Irrigation Systems:** Israel is a global leader in water-efficient agricultural practices, particularly in the use of drip irrigation. This technology has transformed Israeli agriculture, enabling high yields with minimal water use. The success of drip irrigation in Israel serves as a model for water-scarce regions worldwide.
- **Bangladesh's Flood-Resilient Agriculture:** Bangladesh, prone to frequent flooding, has adopted flood-resilient agricultural practices such as floating gardens and flood-tolerant rice varieties. These innovations allow farmers to maintain crop production during the monsoon season, enhancing food security and livelihoods.
- **Africa's Climate-Smart Agriculture Initiatives:** Various African countries have implemented CSA initiatives to enhance agricultural resilience. For example, Kenya's Climate-Smart Agriculture Strategy promotes practices such as agro-forestry, conservation agriculture, and water harvesting. These initiatives have improved food security and farmer incomes while reducing vulnerability to extreme weather.

Challenges and Future Directions

Despite the progress made in adopting resilient agricultural practices, several challenges remain. These include limited access to technology and resources, lack of awareness and training among farmers, and insufficient policy support. Addressing these challenges requires coordinated efforts from governments, research institutions, and the private sector.

- **Access to Technology and Resources:** Ensuring that farmers, particularly smallholders, have access to resilient technologies and resources is crucial. This includes providing affordable seeds, irrigation equipment, and training on resilient farming practices. Governments and development organizations can play a key role in facilitating access through subsidies, grants, and extension services.

- **Policy Support and Incentives:** Developing and implementing policies that support resilient agriculture is essential. This includes providing financial incentives for adopting resilient practices, investing in agricultural research and development, and creating/enabling environments for innovation. Policies should also promote sustainable land and water management to enhance resilience.
- **Education and Capacity Building:** Raising awareness and building the capacity of farmers to adopt resilient practices is critical. This involves providing training programmes, extension services, and access to information on climate risks and adaptive strategies. Farmer cooperatives and networks can facilitate knowledge sharing and collective action for resilience.
- **Research and Innovation:** Continued research and innovation are necessary to develop new resilient technologies and practices. This includes breeding new crop varieties, improving irrigation technologies, and exploring innovative approaches to soil and water management. Collaboration between research institutions, the private sector, and farmers is essential for translating research into practical solutions.
- **International Cooperation and Funding:** Addressing the global challenge of agricultural resilience requires international cooperation and funding. This includes supporting developing countries through financial assistance, technology transfer, and capacity building. International organizations, such as the Food and Agriculture Organization (FAO) and the World Bank,

play a vital role in facilitating global efforts to enhance agricultural resilience.

Conclusion

Agriculture's vulnerability to extreme weather events poses significant risks to food security and economic stability. However, by adopting resilient farming practices and innovative technologies, the agricultural sector can mitigate these risks and enhance its resilience. Strategies such as developing drought-resistant crops, implementing efficient irrigation systems, and building robust flood management infrastructure are essential for sustaining agricultural productivity in the face of climate change. Addressing the challenges of access to technology, policy support, education, research, and international cooperation will be crucial for building a resilient agricultural sector capable of withstanding extreme weather events and ensuring global food security.

Infrastructure: Vulnerability and Resilience to Extreme Weather

Extreme weather events pose significant risks to infrastructure, leading to widespread damage and substantial economic costs. Hurricanes, floods, wildfires, and other extreme weather conditions can compromise the integrity of roads, bridges, buildings, and critical facilities. The aftermath of such events often necessitates extensive repair and recovery efforts, underscoring the importance of investing in resilient infrastructure.

Impact of Extreme Weather on Infrastructure

- **Hurricanes and Coastal Storms:** Hurricanes and coastal storms bring a combination of high winds, heavy rainfall, and storm surges that can devastate infrastructure. For instance, Hurricane Sandy in 2012 caused extensive damage to New York City's infrastructure. The storm flooded subway systems, damaged roads and bridges, and knocked out power to millions of residents. The total economic impact was estimated at $65 billion, with significant portions allocated to infrastructure repair and recovery. Such events highlight the vulnerability of urban infrastructure to severe weather and the pressing need for enhanced resilience.

- **Floods:** Flooding is another major threat to infrastructure, particularly in low-lying and coastal areas. Floodwaters can erode roads, undermine bridges, and damage buildings. For example, the 2010 floods in Pakistan inundated large areas, destroying roads and bridges, and cutting off access to critical services. The economic damage was estimated at $43 billion, illustrating the severe impact of flooding on infrastructure. Ensuring that infrastructure can withstand such events requires careful planning and investment in flood defences and resilient construction practices.
- **Wildfires:** Wildfires, exacerbated by climate change, have become more frequent and intense, posing significant risks to infrastructure. They can destroy homes, businesses, and utility lines, causing widespread disruption. The 2018 Camp Fire in California, for instance, destroyed over 18,000 structures, including homes, schools, and commercial buildings. The economic impact exceeded $16 billion, with substantial costs associated with rebuilding infrastructure. This underscores the importance of fire-resistant building materials and strategic urban planning to mitigate wildfire risks.

Strategies for Resilient Infrastructure

- **Investing in Durable Materials:** One of the key strategies for enhancing infrastructure resilience is the use of durable materials that can withstand extreme weather conditions. For example, building roads and bridges with high-performance concrete and steel can improve their durability and reduce the need for frequent repairs. Similarly, using fire-resistant materials in buildings can mitigate the damage caused by wildfires. Investing in advanced materials not only enhances the longevity of infrastructure but also reduces long-term maintenance costs.

- **Elevated Structures:** Elevating structures, particularly in flood-prone areas, is an effective way to mitigate the impact of extreme weather events. Elevating homes, buildings, and critical infrastructure such as electrical substations can prevent floodwaters from causing extensive damage. For example, after Hurricane Katrina, many buildings in New Orleans were elevated to reduce future flood risks. Elevated infrastructure is particularly important in coastal cities facing rising sea levels and increased flooding.
- **Improved Drainage Systems:** Effective drainage systems are crucial for managing heavy rainfall and preventing flooding. Urban areas need robust storm water management systems that can quickly and efficiently channel water away from streets and buildings. For instance, Tokyo has invested in an extensive underground floodwater diversion system, known as the G-Cans Project, which helps manage and mitigate flood risks.

  Improved drainage infrastructure not only protects urban areas from flooding but also ensures the continued functioning of critical services during extreme weather.
- **Smart Infrastructure:** Integrating smart technologies into infrastructure can enhance its resilience to extreme weather. Smart sensors and monitoring systems can provide real-time data on the condition of infrastructure, allowing for timely maintenance and repairs. For example, smart sensors embedded in bridges can monitor structural integrity and detect potential weaknesses before they lead to catastrophic failures. Similarly, smart grids can enhance the resilience of energy infrastructure by automatically rerouting power during outages caused by extreme weather.
- **Green Infrastructure:** Incorporating green infrastructure, such as green roofs, permeable pavements, and urban

green spaces, can enhance resilience to extreme weather. Green infrastructure can absorb excess rainfall, reduce the urban heat island effect, and provide natural buffers against floods. For instance, New York City has implemented green infrastructure projects to manage storm water and reduce flood risks. These projects not only enhance urban resilience but also improve environmental quality and provide recreational spaces for residents.

- **Policy and Planning:** Effective policy and planning are essential for building resilient infrastructure. Governments must develop and enforce building codes and standards that account for the risks posed by extreme weather. For example, after the devastating impacts of Hurricane Andrew in 1992, Florida implemented stricter building codes to enhance the resilience of new construction to hurricanes. Comprehensive urban planning that considers climate risks can also guide the development of resilient infrastructure and reduce vulnerability.

Conclusion

Extreme weather events present significant challenges to infrastructure, leading to widespread damage and economic losses. Ensuring that infrastructure is resilient to these events involves investing in durable materials, elevated structures, improved drainage systems, and smart and green infrastructure. Effective policy and planning are also crucial for enhancing resilience. By adopting these strategies, communities can protect their infrastructure, reduce recovery costs, and improve their overall resilience to the increasing threat of extreme weather events.

Tourism: Vulnerability to Extreme Weather and Strategies for Resilience

The tourism industry is particularly sensitive to weather conditions, and extreme weather events can have devastating impacts on tourism-dependent economies. Natural disasters such as hurricanes, tsunamis, floods, and wildfires can deter tourists, leading to significant revenue losses and long-term economic repercussions. The vulnerability of the tourism sector to these events underscores the need for strategic planning, diversification of tourism offerings, and enhanced disaster preparedness.

Impact of Extreme Weather on Tourism

- **Natural Disasters and Tourism Decline:** Extreme weather events can cause immediate and significant declines in tourist arrivals. For instance, the 2004 Indian Ocean tsunami had a catastrophic impact on coastal resorts in countries like Thailand, Indonesia, and Sri Lanka. The disaster resulted in widespread destruction of infrastructure, including hotels, restaurants, and tourist attractions, leading to a sharp decline in visitor numbers. The economic impact was profound, with many local economies suffering substantial revenue losses. In Thailand alone, tourism revenue dropped by over $1 billion in the months following the tsunami. Recovery was slow, as rebuilding efforts took time and potential visitors were deterred by the lingering perception of risk.
- **Hurricanes and Coastal Tourism:** Hurricanes pose a significant threat to coastal tourism destinations. For example, Hurricane Irma in 2017 caused extensive damage to popular tourist spots in the Caribbean, including Puerto Rico, the British Virgin Islands, and the Bahamas. The storm destroyed hotels, airports, and marinas, disrupting travel plans and leading to a sharp decline in tourist arrivals. The economic losses were substantial, as tourism is a major source of income for these islands. The recovery process involved not only rebuilding infrastructure but also restoring the confidence of tourists in the safety of these destinations.
- **Floods and Urban Tourism:** Urban tourism can also be severely affected by extreme weather events such as floods.

For instance, Venice, Italy, frequently experiences flooding known as acqua alta, which disrupts daily life and deters tourists. Floodwaters can damage historical sites, hotels, and transportation systems, leading to cancellations and reduced tourist spending. The economic impact extends beyond immediate revenue losses, affecting the livelihoods of those who depend on tourism-related activities.

Strategies for Resilience and Diversification

- **Diversifying Tourism Offerings:** Diversifying tourism offerings is a key strategy to mitigate the impacts of extreme weather events. By developing a range of attractions and activities, destinations can reduce their dependence on a single type of tourism and attract visitors year-round. For example, coastal regions can promote inland attractions such as cultural sites, nature reserves, and adventure tourism. This diversification helps maintain a steady flow of tourists even when coastal areas are affected by extreme weather.
- **Enhancing Disaster Preparedness:** Enhancing disaster preparedness is crucial for protecting tourists and minimizing economic losses. This involves developing and implementing comprehensive disaster management plans that include early warning systems, evacuation procedures, and emergency response protocols. For instance, hotels and resorts in hurricane-prone areas can establish clear evacuation routes and provide guests with information on safety measures. Regular training and drills for staff ensure that everyone knows their roles in the event of a disaster. Effective communication with tourists before and during extreme weather events can also help manage their expectations and ensure their safety.
- **Building Resilient Infrastructure:** Investing in resilient infrastructure is essential for minimizing the damage caused by extreme weather events. This includes constructing

buildings that can withstand high winds, flooding, and other weather-related hazards. For example, hotels in hurricane-prone areas can be built with reinforced structures and elevated to avoid floodwaters. Additionally, implementing green infrastructure, such as rain gardens and permeable pavements, can help manage storm water and reduce the risk of flooding.

- **Promoting Sustainable Tourism:** Sustainable tourism practices can enhance the resilience of tourism destinations to extreme weather events. This involves adopting environmentally friendly practices that reduce the impact of tourism on natural resources and ecosystems. For instance, eco-tourism initiatives that focus on conservation and sustainable use of natural areas can attract visitors who are interested in responsible travel.
- **Leveraging Technology and Data:** Technology and data can play a significant role in enhancing the resilience of the tourism sector. Advanced weather forecasting systems can provide timely warnings of extreme weather events, allowing destinations to take proactive measures to protect tourists and infrastructure. Additionally, data analytics can help identify trends and patterns in tourist behaviour, enabling destinations to develop targeted marketing strategies and adjust their offerings to changing conditions.
- **Collaboration and Partnerships:** Collaboration between the public and private sectors is essential for building resilience in the tourism industry. Governments, tourism boards, and private businesses can work together to develop and implement disaster preparedness plans, share best practices, and pool resources for recovery efforts. Public-Private Partnerships can also support the development of resilient infrastructure and the promotion of sustainable tourism practices.
- **Insurance and Financial Mechanisms:** Insurance can provide financial protection against the economic losses caused by extreme weather events. Destinations can invest

in insurance policies that cover property damage, business interruption, and other losses related to natural disasters. Additionally, establishing emergency funds and access to financial assistance can help businesses recover more quickly after a disaster.

Conclusion

The tourism industry is highly sensitive to extreme weather events, which can cause significant economic losses and long-term disruptions. However, by diversifying tourism offerings, enhancing disaster preparedness, building resilient infrastructure, promoting sustainable tourism, leveraging technology and data, and fostering collaboration and partnerships, destinations can mitigate these impacts and build resilience. Effective strategies for resilience not only protect the tourism sector from the immediate effects of extreme weather but also ensure its long-term sustainability and success. Through proactive measures and a commitment to sustainability, the tourism industry can navigate the challenges posed by extreme weather and continue to thrive in the face of climate change.

Energy: Vulnerability to Extreme Weather and Strategies for Resilience

Extreme weather events can significantly disrupt energy production and distribution, leading to widespread power outages, increased energy costs, and heightened reliance on fossil fuels. The 2011 Japanese tsunami serves as a stark reminder of the vulnerability of energy infrastructure to natural disasters. Building a resilient energy infrastructure that incorporates renewable energy sources and robust distribution networks is essential for maintaining energy security during extreme weather events.

Impact of Extreme Weather on Energy Infrastructure

- **The 2011 Japanese Tsunami:** The earthquake and subsequent tsunami that struck Japan in 2011 caused severe damage to the Fukushima Daiichi nuclear power plant, leading to the shutdown of several reactors. This disaster not only resulted in immediate energy shortages but also had long-term implications for Japan's energy policy. The country had to increase its reliance on fossil fuels, particularly liquefied natural gas (LNG), to compensate for the loss of nuclear power, leading to higher greenhouse gas emissions and increased energy costs. The Fukushima disaster highlighted the need for resilient energy infrastructure capable of withstanding extreme natural events.

- **Hurricanes and Storms:** Hurricanes and storms can cause extensive damage to power lines, substations, and generation facilities. For instance, Hurricane Maria in 2017 devastated Puerto Rico's electrical grid, resulting in prolonged power outages that lasted for months in some areas. The storm destroyed transmission lines, damaged power plants, and crippled the island's energy distribution network, severely impacting the daily lives of residents and hindering economic recovery.
- **Heatwaves and Wildfires:** Heatwaves can strain energy systems by increasing the demand for air conditioning and cooling, leading to potential overloads and blackouts. Wildfires, on the other hand, can damage transmission lines and substations, disrupting electricity supply.

  The 2019-20 Australian bushfires, for example, caused significant damage to the energy infrastructure, leading to power outages and increased operational costs for utilities.

Strategies for Building Resilient Energy Infrastructure

- **Diversifying Energy Sources:** Diversifying energy sources is a crucial strategy for enhancing resilience. Relying on a single energy source can make a system vulnerable to disruptions. Incorporating a mix of renewable energy sources, such as solar, wind, hydro, and geothermal, can reduce dependency on any one source and enhance overall energy security. For example, Denmark's energy strategy includes a diverse mix of wind, solar, and bioenergy, making its energy system more resilient to extreme weather.

- **Investing in Renewable Energy:** Renewable energy sources are less susceptible to the impacts of extreme weather compared to centralized fossil fuel and nuclear power plants. Distributed renewable energy systems, such as rooftop solar panels and small-scale wind turbines, can continue to operate independently even if parts of the grid are damaged. Investing in renewable energy not only reduces greenhouse gas emissions but also enhances the resilience of the energy infrastructure.
- **Robust Distribution Networks:** Strengthening the energy distribution network is essential for ensuring reliable power supply during extreme weather events. This includes upgrading transmission lines, reinforcing substations, and implementing smart grid technologies. Smart grids can detect and respond to disruptions in real-time, automatically rerouting power to avoid outages and minimize damage. For instance, Italy's Enel has implemented a smart grid system that enhances the resilience and efficiency of its electricity distribution network.
- **Micro Grids and Energy Storage:** Developing micro grids and energy storage systems can enhance energy resilience by providing localized sources of power that can operate independently of the main grid. Micro grids can integrate various energy sources, including renewables and storage, to supply power during grid outages. Energy storage systems, such as batteries, can store excess renewable energy and release it when needed, ensuring a continuous power supply. California has been a leader in adopting micro grids and energy storage solutions to enhance the resilience of its

energy infrastructure against wildfires and other extreme weather events.

- **Hardening Infrastructure:** Hardening energy infrastructure involves making physical improvements to withstand extreme weather conditions. This can include elevating power substations above flood levels, reinforcing power lines to withstand high winds, and installing fire-resistant materials around critical infrastructure. For example, Florida Power & Light has invested heavily in hardening its grid infrastructure; including replacing wooden poles with concrete ones and elevating substations, to better withstand hurricanes.
- **Policy and Regulatory Support:** Effective policies and regulations are essential for promoting resilient energy infrastructure. Governments can provide incentives for renewable energy adoption, invest in grid modernization, and implement regulations that require energy companies to enhance the resilience of their systems. For instance, the European Union's Clean Energy for All Europeans package includes measures to promote renewable energy, improve energy efficiency, and enhance grid resilience.
- **Community Engagement and Preparedness:** Engaging communities in energy resilience planning is crucial. Educating the public about energy-saving practices, emergency preparedness, and the benefits of renewable energy can foster community support for resilience initiatives. Community-based energy projects, such as local solar farms and cooperative wind projects, can enhance local energy security and resilience. Additionally, involving communities in the planning and implementation of resilience projects ensures that local needs and preferences are considered.
- **Research and Innovation:** Ongoing research and innovation are vital for developing new technologies and strategies to enhance energy resilience. This includes

advancements in renewable energy technologies, smart grid systems, and energy storage solutions. Collaborative research efforts between governments, academia, and the private sector can drive innovation and accelerate the adoption of resilient energy technologies.

Conclusion

Extreme weather events pose significant challenges to energy infrastructure, leading to disruptions in energy production and distribution. Building a resilient energy infrastructure that includes diverse renewable energy sources, robust distribution networks, micro grids, and energy storage systems is crucial for maintaining energy security during extreme weather. Investing in infrastructure hardening, smart grid technologies, and community engagement, supported by effective policies and regulations, can further enhance resilience.

Through proactive measures and a commitment to innovation, the energy sector can better withstand the impacts of extreme weather, ensuring a reliable and sustainable energy supply for the future.

Insurance: Adapting to the Growing Frequency and Severity of Extreme Weather Events

The insurance sector is increasingly challenged by the financial impact of extreme weather events, which are becoming more frequent and severe due to climate change. These events lead to higher claims and substantial financial losses for insurers, necessitating adjustments in risk models, pricing strategies, and the development of innovative insurance products to ensure the industry's resilience and sustainability.

Impact of Extreme Weather on the Insurance Sector

- **Increased Claims and Financial Losses:** Extreme weather events such as hurricanes, floods, wildfires, and storms result in significant property damage and loss, leading to a surge in insurance claims. For example, the 2017 Atlantic hurricane season, which included Hurricanes Harvey, Irma, and Maria, caused record-breaking insured losses exceeding $100 billion. This trend of escalating claims underscores the need for insurers to adapt their strategies to manage the increasing financial risks associated with extreme weather.

- **Higher Premiums and Stricter Underwriting:** In response to the growing risks, insurers have been compelled to raise premiums and implement stricter underwriting standards, particularly for properties in vulnerable areas. For instance, properties located in hurricane-prone coastal regions or areas with high flood risk face higher insurance costs and more rigorous coverage requirements. This approach helps insurers manage risk exposure but can also lead to affordability issues for policyholders, potentially resulting in underinsurance or lack of coverage.

Adjusting Risk Models and Pricing Strategies

- **Enhanced Risk Assessment:** To accurately price insurance products and manage risk exposure, insurers must refine their risk assessment models.
  This involves incorporating the latest climate science data, historical loss records, and advanced predictive analytics to forecast the potential impacts of extreme weather events. By leveraging technology and data analytics, insurers can improve their understanding of risk patterns and better anticipate future losses.
- **Dynamic Pricing:** Implementing dynamic pricing strategies allows insurers to adjust premiums in real-time based on evolving risk conditions. For example, insurers can use weather forecasting and real-time monitoring tools to assess imminent threats and adjust premiums accordingly. This proactive approach helps insurers manage risk more effectively and ensures that premiums reflect the current level of exposure.

- **Reinsurance and Risk Transfer:** Reinsurance plays a critical role in managing the financial impact of extreme weather events. By transferring a portion of their risk to reinsurance companies, primary insurers can mitigate their exposure to large-scale losses. Additionally, innovative risk transfer mechanisms, such as catastrophe bonds (cat bonds), allow insurers to securitize risk and attract investment from capital markets. These financial instruments provide an additional layer of protection and help insurers manage their capital more efficiently.

Innovative Insurance Products

- **Parametric Insurance:** Parametric insurance is an innovative product that offers a more efficient risk transfer mechanism compared to traditional indemnity-based insurance. Instead of indemnifying the actual loss, parametric insurance provides a predefined payout based on the occurrence of a specific event, such as a hurricane of a certain category or a specific amount of rainfall. This approach ensures quicker payouts, reducing the financial burden on policyholders and speeding up recovery efforts. For example, the Caribbean Catastrophe Risk Insurance Facility (CCRIF) uses parametric insurance to provide immediate financial support to Caribbean governments following extreme weather events.
- **Index-based Insurance:** Similar to parametric insurance, index-based insurance relies on objective indices, such as rainfall levels or temperature thresholds, to trigger payouts. This product is particularly beneficial for sectors like agriculture, where losses can be directly correlated with weather conditions. Index-based insurance reduces the complexity of claims processing and provides timely financial relief to affected farmers, enhancing their resilience to extreme weather.

- **Climate Risk Insurance Pools:** Establishing insurance pools can spread the risk of extreme weather events across a larger group of insurers and policyholders. These pools can provide coverage for high-risk areas and ensure that premiums remain affordable.
  For example, the Flood Re scheme in the United Kingdom is a joint initiative between the government and insurers to provide affordable flood insurance to properties at high risk of flooding. By pooling resources and spreading risk, such initiatives enhance the availability and affordability of coverage.
- **Micro Insurance:** Micro insurance products are designed to provide affordable coverage to low-income individuals and communities that are particularly vulnerable to extreme weather events. These products typically offer lower coverage limits and premiums, making them accessible to those who might otherwise be uninsured. Micro insurance can play a crucial role in enhancing the resilience of vulnerable populations by providing financial protection and support in the aftermath of disasters.

Future Directions and Challenges

- **Regulatory Environment:** The insurance sector operates within a complex regulatory framework that varies by region. Regulators must balance the need for consumer protection with the sustainability of the insurance industry. This includes setting solvency requirements, ensuring fair pricing practices, and encouraging innovation in product offerings. Governments and regulatory bodies can also promote resilience by incentivizing risk reduction measures, such as retrofitting buildings and investing in resilient infrastructure.
- **Consumer Education and Engagement:** Educating consumers about the importance of insurance and the availability of innovative products is essential for increasing

coverage and reducing underinsurance. Insurers can engage with policyholders through awareness campaigns, educational programmes, and transparent communication about risk and coverage options.

- **Collaboration and Data Sharing:** Collaboration between insurers, governments, and other stakeholders is crucial for developing effective risk management strategies. Sharing data on climate risks, loss patterns, and mitigation efforts can enhance risk assessment models and improve the accuracy of pricing. Public-private partnerships can also facilitate the development of innovative products and solutions that address the evolving challenges posed by extreme weather.
- **Investment in Resilience:** Beyond providing financial protection, insurers can play a proactive role in promoting resilience. This includes investing in risk reduction measures, supporting climate adaptation projects, and advocating for policies that address the underlying causes of climate change. By aligning their investment strategies with resilience goals, insurers can contribute to broader efforts to mitigate the impacts of extreme weather and enhance the sustainability of communities.

Conclusion

The insurance sector faces significant challenges in managing the financial risks associated with the increasing frequency and severity of extreme weather events. Adapting risk models, pricing strategies, and product offerings is essential for maintaining the industry's resilience and sustainability. Innovative insurance products, such as parametric and index-based insurance, provide efficient risk transfer mechanisms and timely financial relief to policyholders. Collaboration, regulatory support, and consumer engagement are critical for enhancing the effectiveness and reach of insurance coverage. By investing in resilience and promoting sustainable practices, the insurance sector can play a vital role in mitigating the economic impacts of extreme weather and supporting the long-term stability of communities.

Preparedness and Response Strategies

Effective preparedness and response strategies are essential for minimizing the economic impacts of extreme weather events.

Early Warning Systems: Critical Tools for Saving Lives and Reducing Economic Losses

Early Warning Systems (EWS) are essential tools for mitigating the impacts of extreme weather events. By providing timely and accurate information about impending natural disasters, these systems enable communities, governments, and businesses to take proactive measures to protect lives, property, and economic assets. Effective early warning systems incorporate advanced meteorological technologies, robust communication networks, and comprehensive preparedness plans to enhance resilience against hurricanes, floods, storms, and other extreme weather conditions.

Importance of Early Warning Systems

- *Saving Lives:* The primary objective of early warning systems is to save lives by providing advance notice of extreme weather events. When communities receive timely alerts about imminent dangers, they can evacuate or seek shelter, significantly reducing the risk of casualties. For instance, during Hurricane Katrina in 2005, despite the devastation, the early warning system played a critical role in saving thousands of lives by prompting mass evacuations.
- *Reducing Economic Losses:* Early warning systems also help reduce economic losses by allowing individuals and organizations to take protective measures before a disaster strikes. This can include securing property, moving assets to safer locations, and implementing business continuity plans. For example, advanced warnings of Hurricane Sandy in 2012 enabled businesses and residents in New York City

to prepare, mitigating some of the economic impacts of the storm.

- *Enhancing Community Resilience:* Effective early warning systems strengthen community resilience by fostering a culture of preparedness and proactive risk management. Communities, who are well-informed about potential hazards and know how to respond are better equipped to recover quickly from disasters. This resilience is crucial for maintaining social and economic stability in the face of increasing climate risks.

Components of Effective Early Warning Systems

- *Advanced Meteorological Technologies:* Accurate forecasting is the backbone of any early warning system. Advanced meteorological technologies, such as satellite imaging, radar, and weather modelling, provide precise and timely information about developing weather patterns. For instance, the use of Doppler radar allows meteorologists to track storm systems and predict their movement with high accuracy.
  Integrating these technologies into early warning systems ensures that alerts are based on reliable data and forecasts.
- *Robust Communication Networks:* Effective communication is vital for disseminating warnings to the public and relevant stakeholders. Early warning systems must incorporate multiple communication channels, including television, radio, mobile alerts, social media, and sirens, to reach diverse audiences. For example, Japan's nationwide earthquake early warning system uses a combination of mobile alerts, TV broadcasts, and public announcement systems to notify residents of impending seismic activity. Ensuring that warnings are accessible and comprehensible to all segments of the population, including those with disabilities and language barriers, is crucial for maximizing the effectiveness of early warnings.

- *Community Engagement and Education:* Public awareness and education are critical components of early warning systems. Communities need to understand the risks they face, recognize warning signals, and know how to respond appropriately. Education campaigns, community drills, and training programmes can enhance public knowledge and preparedness. For example, the "Ready New York" campaign provides residents with information on how to prepare for various emergencies, including hurricanes and floods, through brochures, workshops, and online resources.
- *Integrated Response Plans:* Early warning systems should be integrated with comprehensive response plans that outline specific actions to be taken during different types of emergencies. These plans should involve coordination between local authorities, emergency services, healthcare providers, and community organizations. For instance, the National Hurricane Centre in the United States works closely with federal, state, and local agencies to develop and implement response strategies during hurricane season. Effective coordination ensures a swift and organized response, minimizing confusion and enhancing the efficiency of evacuation and relief efforts.
- *International Collaboration and Data Sharing:* Extreme weather events often have transboundary impacts, necessitating international cooperation and data sharing. Collaborative efforts, such as the Global Framework for Climate Services (GFCS) and the World Meteorological Organization (WMO), facilitate the exchange of weather data and best practices among countries.
- *Technological Innovation and Adaptation:* Continual innovation and adaptation of technologies are essential for improving early warning systems. Advances in artificial intelligence (AI) and machine learning (ML) can enhance predictive capabilities by analyzing vast amounts of data

and identifying patterns that traditional methods might miss. Additionally, integrating Internet of Things (IoT) devices, such as weather sensors and smart infrastructure, can provide real-time data and automate warning dissemination. For example, AI-driven flood forecasting systems can predict river flooding with greater precision, enabling timely evacuation and resource allocation.

Case Studies

- **Bangladesh's Cyclone Preparedness Programme:** Bangladesh has developed an effective early warning system for cyclones, combining meteorological forecasting with community-based preparedness. The Cyclone Preparedness Programme (CPP) uses a network of volunteers to disseminate warnings and assist with evacuations. This programme has significantly reduced cyclone-related fatalities, as seen during Cyclone Sidr in 2007, where early warnings and efficient evacuations saved countless lives.
- **Mexico's Earthquake Early Warning System:** Mexico's Sistema de Alerta Sísmica Mexicano (SASMEX) is an advanced earthquake early warning system that provides alerts to residents in high-risk areas. The system uses a network of seismic sensors to detect earthquakes and issue warnings *via* television, radio, and mobile alerts. During the 2017 Puebla earthquake, SASMEX provided crucial seconds of warning, allowing people to take protective actions and reducing casualties.
- **The Pacific Tsunami Warning Centre:** The Pacific Tsunami Warning Centre (PTWC) provides tsunami warnings to countries in the Pacific Basin. Using seismic data and ocean buoys, the PTWC monitors and predicts tsunamis, issuing alerts to vulnerable coastal communities. The system's effectiveness was demonstrated during the 2011 T-hoku earthquake and tsunami, where early warnings allowed for timely evacuations in several countries.

Challenges and Future Directions

- **Accessibility and Inclusivity:** Ensuring that early warning systems are accessible to all individuals, including those with disabilities and language barriers, remains a challenge. Tailoring warning messages to meet the needs of diverse populations and ensuring that communication channels are inclusive is crucial for maximizing the reach and effectiveness of alerts.
- **Sustainability and Funding:** Sustaining and funding early warning systems, particularly in developing countries, can be challenging. Continuous investment in technology, infrastructure, and community engagement is necessary to maintain and improve these systems. International aid and partnerships can play a vital role in supporting early warning systems in resource-constrained regions.
- **Climate Change and Evolving Risks:** Climate change is altering the frequency, intensity, and patterns of extreme weather events, necessitating continual adaptation of early warning systems. Integrating climate projections and emerging risks into early warning frameworks ensure that systems remain relevant and effective in the face of changing climate conditions.

Conclusion

Early warning systems are critical tools for mitigating the impacts of extreme weather events, saving lives, and reducing economic losses. By leveraging advanced meteorological technologies, robust communication networks, and comprehensive preparedness plans, communities can enhance their resilience to natural disasters. Continued innovation, international collaboration, and inclusive strategies are essential for improving the effectiveness of early warning systems and ensuring that all individuals are protected from the growing threats posed by extreme weather events. Through proactive measures and sustained investment, early warning systems can play a pivotal role in safeguarding communities and supporting sustainable development.

REFERENCES

Gali, J. (2022). *Monetary Policy, Inflation, and the Business Cycle: An Introduction to the New Keynesian Framework*, 3rd edn., Princeton University Press.

Gopinath, G. (2022). "The International Dimension of Inflation: Evidence and Policy Implications". *IMF Staff Papers, 69*(2): 189-218.

Gorton, G. and Metrick, A. (2022). "The Fed and the Great Inflation of the 1970s: The Evidence Speaks". *Journal of Monetary Economics, 124*: 94-114.

Guerrieri, V., Lorenzoni, G., Straub, L. and Werning, I. (2022). "Macroeconomic Implications of COVID-19: Can Negative Supply Shocks Cause Demand Shortages?" *American Economic Review: Insights, 4*(2): 173-190.

Guvenen, F., Kaplan, G. and Song, J. (2022). "How Does Inflation Affect Different Income Groups?" *Journal of Political Economy, 130*(5): 1437-1473.

Ha, J., Kose, M.A. and Ohnsorge, F. (2022). "Global Inflation Synchronization". *Journal of International Economics*, Vol. 134, Editorial Board Article No. 103595.

Hall, R.E. (2022). "The Limits of Inflation Targeting". *Journal of Economic Perspectives, 36*(2): 3-24.

Hamilton, J.D. (2022). "Why You Should Never Use the Hodrick-Prescott Filter". *Review of Economic Studies, 89*(1): 104-120.

Jordà, Ò. and Taylor, A.M. (2022). "The Pandemic, Inflation, and Debt Sustainability". *Economic Policy, 37*(110): 275-314.

Jordà, Ò., Singh, S.R. and Taylor, A.M. (2022). "The Long-Run Effects of Monetary Policy". *Journal of Monetary Economics, 123*: 1-20.

Kamin, S.B. (2022). "Global Supply Chains in a Post-Pandemic World: Economic Insights and Policy Responses". *International Journal of Central Banking, 18*(3): 57-88.

Kiley, M.T. (2022). "Unemployment, Inflation, and Monetary Policy: Evidence from Advanced and Emerging Market Economies". *Journal of International Money and Finance*, Vol. 122, Editorial Board Article No. 102564.

Krugman, P. (2022). "Inflation and the Power of Economic Ideas". *Foreign Affairs, 101*(2): 23-37.

Lane, P.R. (2022). "Global Supply Chains and Inflation: The Role of the Pandemic and Policy Responses". *IMF Economic Review, 70*(2): 121-143.

Laubach, T. and Williams, J.C. (2022). "Measuring the Natural Rate of Interest: International Trends and Determinants". *Brookings Papers on Economic Activity, 2022*(3): 57-108.

Leduc, S. and Wilson, D. (2022). "What is Driving the Rise in Inflation?: A Primer". *Economic Letter*, Federal Reserve Bank of San Francisco, pp. 2022-27.

Mishkin, F.S. (2022). "Monetary Policy Strategy: Lessons from the Financial Crisis". *Journal of Economic Perspectives, 36*(3): 123-148.

Mussa, M. (2022). "Inflation, Real Interest Rates, and the Real Exchange Rate: An Empirical Investigation". *Journal of Monetary Economics, 127:* 243-271.

Orphanides, A. (2022). "Policy Rules and Economic Stability: Lessons from the Great Inflation". *Journal of Economic Dynamics and Control*, Vol. 135, Editorial Board Article No. 104313.

Reis, R. (2022). "The Analytics of Monetary Policy: A New Keynesian Perspective". *Quarterly Journal of Economics, 137*(4): 1951-1987.

Sims, C.A. (2022). "Fiscal Policy, Inflation, and Debt Sustainability". *Journal of Economic Theory*, Vol. 198, Editorial Board Article No. 105297.

Smets, F. and Wouters, R. (2022). "An Estimated Dynamic Stochastic General Equilibrium Model of the Euro Area". *Journal of Economic Dynamics and Control*, Vol. 133, Editorial Board Article No. 104295.

Stock, J.H. and Watson, M.W. (2022). "Disentangling the Channels of the 2021-2022 Inflation Surge". *Journal of Applied Econometrics, 37*(6): 975-996.

10

Sustainable Business Practices

Aligning Profit with Planet

Role of Businesses in Climate Adaptation

Businesses play a pivotal role in climate adaptation by integrating sustainable practices into their operations and supply chains. As major consumers of resources and significant contributors to environmental change, businesses have both the responsibility and the capability to drive adaptation efforts. Their involvement is crucial in reducing carbon footprints, promoting resource efficiency, and fostering innovation in climate resilience.

Leadership in Sustainability

Setting Ambitious Sustainability Goals: Leading companies understand that sustainability is not just a trend but a critical component of long-term success. By setting ambitious sustainability goals, businesses can drive significant changes within their operations and beyond. These goals often include targets for reducing carbon footprints, enhancing resource efficiency, and achieving zero waste. For example, global retailer Walmart has committed to achieving zero emissions across its global operations by 2040, including powering its facilities with 100 per cent renewable energy by 2035. Such bold commitments set a high standard and signal to the industry that sustainable practices are essential for future competitiveness.

- **Reducing Greenhouse Gas Emissions:** One of the most impactful ways companies can lead in sustainability is by actively working to reduce their greenhouse gas emissions. This involves implementing energy efficiency measures, optimizing supply chains, and transitioning to low-carbon technologies. Tech giant Apple, for instance, has made significant strides in reducing its carbon footprint by ensuring that all of its facilities worldwide are powered by 100 per cent renewable energy. Furthermore, Apple is working towards making its entire supply chain carbon-neutral by 2030. By addressing emissions comprehensively, companies can significantly contribute to global efforts to combat climate change.
- **Investing in Renewable Energy:** Transitioning to renewable energy sources is a key strategy for companies aiming to lead in sustainability.
  Investments in solar, wind and other renewable energy technologies not only reduces a company's reliance on fossil fuels but also promotes the broader adoption of clean energy. Google, a pioneer in this area, has been the largest corporate purchaser of renewable energy for several years. The company has invested in large-scale renewable energy projects, such as wind farms and solar plants, to power its data centres and offices. These investments not only support Google's sustainability goals but also stimulate growth and innovation in the renewable energy sector.
- **Influencing Industry Practices:** Companies that lead in sustainability often have a ripple effect across their industries, encouraging other businesses to adopt similar practices. This influence can be exerted through various means, including industry coalitions, supply chain

standards, and public advocacy. For example, the Fashion Industry Charter for Climate Action, initiated by the United Nations, brings together leading fashion brands to commit to reducing greenhouse gas emissions. Companies like Nike and H&M, which are part of this charter, use their market influence to promote sustainable practices throughout their supply chains, encouraging suppliers and competitors alike to follow suit.

- **Public Reporting and Transparency:** Transparency in reporting sustainability efforts is crucial for building trust and accountability. Leading companies publish detailed sustainability reports that outline their goals, progress, and challenges. These reports often follow internationally recognized frameworks, such as the Global Reporting Initiative (GRI) or the Task Force on Climate-related Financial Disclosures (TCFD). By openly sharing their sustainability strategies and outcomes, companies like Unilever and Patagonia not only demonstrate their commitment to sustainability but also provide valuable insights and benchmarks for other businesses.
- **Employee Engagement and Culture:** Companies that lead in sustainability often embed these values deeply within their organizational culture. This involves engaging employees at all levels, from executives to frontline workers, in sustainability initiatives. Internal programmes that encourage sustainable practices, such as reducing waste, conserving energy, and participating in volunteer activities, help cultivate a culture of sustainability.

  For instance, Salesforce has integrated sustainability into its corporate culture by encouraging employees to participate in green initiatives and providing sustainability education and resources. This holistic approach ensures that sustainability is not just a top-down mandate but a shared value across the organization.

- **Innovation and Research:** Leading companies invest in research and development to drive innovation in sustainable products and practices. By exploring new materials, technologies, and processes, businesses can create solutions that reduce environmental impact and enhance sustainability. Tesla, for example, has revolutionized the automotive industry with its electric vehicles and energy storage solutions. Through continuous innovation, Tesla not only advances its sustainability goals but also sets new standards for the industry, pushing competitors to develop cleaner and more efficient technologies.

Conclusion

Leadership in sustainability involves setting ambitious goals, reducing greenhouse gas emissions, investing in renewable energy, and influencing industry practices. Companies that prioritize sustainability can drive significant changes within their sectors and encourage others to adopt similar practices. Transparency in reporting, employee engagement, and a culture of innovation further enhance a company's ability to lead in sustainability. By integrating these strategies, businesses not only contribute to environmental stewardship but also achieve long-term success and resilience in an increasingly sustainability-focused market.

Innovation and Technology in Climate Resilience

Developing and Deploying New Technology Businesses have the resources and expertise to drive technological innovation, which is crucial for enhancing climate resilience. By investing in research and development (R&D), companies can create and implement new technologies that address environmental challenges and improve adaptive capacities.

Innovations in Renewable Energy

- **Solar and Wind Energy:** Renewable energy technologies, particularly solar and wind, have seen significant advancements due to corporate investment and innovation. Companies like Tesla and Siemens have pioneered in developing efficient and affordable solar panels and wind turbines. These technologies not only reduce dependence on fossil fuels but also provide sustainable energy solutions that can withstand the impacts of climate change. Solar and wind farms are increasingly being integrated into national grids, providing clean energy and reducing greenhouse gas emissions.

- **Energy Storage Solutions:** One of the major challenges of renewable energy is its intermittency. Innovations in energy storage, such as Tesla's Powerwall and Powerpack systems, allow for the storage of excess energy generated during peak production times for use during periods of low production. These advancements ensure a stable and reliable energy supply, enhancing energy security and resilience. Energy storage solutions are particularly crucial for integrating renewable energy into the grid and supporting off-grid applications in remote or underserved areas.

Energy Efficiency Technologies

- **Smart Grids:** Smart grid technology enhances the efficiency and reliability of electricity distribution. By using digital communication technology, smart grids can detect and respond to local changes in energy usage, integrate renewable energy sources, and reduce energy losses. Companies like General Electric and IBM are at the forefront of developing smart grid solutions that enhance the resilience and efficiency of power systems.
- **Building Automation Systems:** Innovations in building automation and energy management systems help reduce energy consumption in commercial and residential buildings. These systems use sensors and advanced algorithms to optimize heating, cooling, lighting, and other energy-intensive operations. Companies like Honeywell

and Johnson Controls provide smart building solutions that enhance energy efficiency and reduce operational costs, contributing to lower carbon footprints and increased resilience to energy disruptions.

Advancements in Waste Management

- **Circular Economy Practices:** Businesses are increasingly adopting circular economy principles, which focus on reducing waste, reusing materials, and recycling products. Innovations in this area include the development of biodegradable materials, advanced recycling technologies, and product life extension strategies. For example, companies like Loop Industries are developing chemical recycling processes that break down plastics into their basic building blocks for reuse in new products. This reduces waste and conserves resources, mitigating environmental impacts and enhancing sustainability.
- **Waste-to-Energy Technologies:** Converting waste into energy is an innovative solution that addresses both waste management and energy generation. Technologies such as anaerobic digestion and gasification can convert organic waste into biogas and syngas, which can be used to generate electricity and heat. Companies like Veolia and Waste Management Inc. are investing in waste-to-energy projects that provide renewable energy while reducing landfill use and greenhouse gas emissions.

Sustainable Agriculture

- **Precision Farming:** Precision farming technologies use data analytics, GPS, and IoT devices to optimize agricultural practices. By monitoring soil conditions, weather patterns, and crop health in real-time, farmers can make informed decisions about irrigation, fertilization, and pest control. Companies like John Deere and Trimble are developing precision farming tools that enhance productivity, reduce resource use, and increase resilience to climate variability.

- **Drought-Resistant Crops:** Biotechnology and genetic engineering have led to the development of drought-resistant crops, which are crucial for maintaining agricultural productivity in the face of changing climate conditions. Companies like Monsanto (now part of Bayer) and Syngenta are investing in research to develop crop varieties that can withstand extreme weather conditions, reducing the risk of crop failure and ensuring food security.
- **Vertical Farming:** Vertical farming is an innovative agricultural practice that involves growing crops in vertically stacked layers, often in controlled indoor environments. This method reduces the need for arable land, minimizes water usage, and protects crops from extreme weather. Companies like Aero Farms and Plenty are leading the way in vertical farming, providing sustainable and resilient food production solutions.

Economic and Environmental Benefits

- **Reducing Environmental Impacts:** Innovations in renewable energy, energy efficiency, waste management, and sustainable agriculture significantly reduce environmental impacts. By decreasing greenhouse gas emissions, conserving resources, and minimizing waste, these technologies contribute to a healthier planet and mitigate the effects of climate change.
- **Enhancing Adaptive Capacities:** Technological advancements enhance the ability of businesses and communities to adapt to changing climate conditions. Improved energy security, resilient infrastructure, and sustainable agricultural practices provide the foundation for enduring climate resilience.
- **Creating Market Opportunities:** Investing in climate resilience technologies opens up new market opportunities

for businesses. The growing demand for sustainable products and services drives innovation and economic growth. Companies that lead in developing and deploying these technologies gain a competitive edge and attract environmentally conscious consumers and investors.

Conclusion

Innovation and technology are crucial for enhancing climate resilience. Businesses play a key role in developing and deploying new technologies in renewable energy, energy efficiency, waste management, and sustainable agriculture. These advancements significantly reduce environmental impacts, improve adaptive capacities, and create economic opportunities. By prioritizing innovation, businesses can drive sustainable development and contribute to a resilient and sustainable future. Through continuous investment in R&D and the adoption of cutting-edge technologies, companies can lead the way in addressing the challenges posed by climate change.

Collaboration and Partnerships

Businesses can significantly enhance their impact on addressing climate challenges by collaborating with governments, non-profits, and other stakeholders. Public-Private Partnerships (PPPs) are particularly effective in mobilizing resources, sharing expertise, and implementing large-scale adaptation projects that benefit both businesses and communities.

The Importance of Collaboration

- **Resource Mobilization:** Climate adaptation and mitigation efforts often require substantial financial and technical resources that may be beyond the capacity of any single entity. Collaborations between businesses, governments, and non-profits can pool resources, enabling the implementation of large-scale projects. For example, the Green Climate Fund, established to support developing countries in responding to climate change, mobilizes resources from both public and private sectors to fund projects aimed at reducing greenhouse gas emissions and enhancing climate resilience.

- **Expertise Sharing:** Each stakeholder in a partnership brings unique skills, knowledge, and perspectives. Governments can provide regulatory frameworks and incentives, businesses can offer technological innovation and operational efficiency, and non-profits can contribute on-the-ground experience and community engagement. This synergy allows for the development of comprehensive and effective climate strategies. For instance, the collaboration between the World Wildlife Fund (WWF) and multinational corporations like Coca Cola focuses on water conservation, leveraging WWF's expertise in conservation and Coca Cola's operational reach.
- **Implementing Large-Scale Projects:** Large-scale adaptation projects often require coordination and cooperation across various sectors and stakeholders. PPPs enable the implementation of projects that might be too complex or costly for individual entities. For example, the Thames Estuary 2100 project in the UK involves multiple stakeholders, including government agencies, engineering firms, and environmental groups, working together to protect London from future flood risks through comprehensive and integrated flood defence strategies.

Examples of Successful Public-Private Partnerships

- **Rockefeller Foundation's 100 Resilient Cities Initiative:** The 100 Resilient Cities initiative, launched by the Rockefeller Foundation, is a prime example of a successful PPP. This initiative supports cities worldwide in becoming more resilient to physical, social, and economic challenges. By partnering with local governments, businesses, and non-profits, the initiative provides cities with resources, tools, and expertise to develop and implement resilience strategies. The collaboration fosters innovative solutions and leverages the strengths of different sectors to enhance urban resilience.

- **Climate Finance Partnerships:** The Climate Finance Partnership, an initiative involving asset managers like Black Rock and public entities like the French Development Agency, aims to catalyze private investment in climate infrastructure projects in emerging markets. By de-risking investments through blended finance models, this partnership mobilizes significant private capital to support sustainable energy, transportation, and water projects, addressing climate change while promoting economic development.
- **Sustainable Agriculture Partnerships:** The Sustainable Agriculture Initiative Platform (SAI) is a global food industry partnership that promotes sustainable agricultural practices. Companies like Nestlé, Unilever, and Danone collaborate with farmers, governments, and non-profits to develop and implement best practices for sustainable agriculture. This partnership enhances food security, reduces environmental impacts, and improves the livelihoods of farmers.

Benefits of Collaboration and Partnerships

- **Enhanced Resilience:** By leveraging the strengths of different stakeholders, PPPs can develop more robust and comprehensive climate adaptation strategies. Collaborative efforts ensure that solutions are well-rounded, addressing various aspects of resilience, from infrastructure to community engagement.
- **Innovation and Efficiency:** Partnerships foster innovation by combining diverse expertise and perspectives. Businesses bring cutting-edge technologies and efficiency-driven

approaches, while governments and non-profits contribute regulatory support and community insights. This collaboration leads to the development of innovative and efficient solutions to climate challenges.

- **Scalability and Impact:** Large-scale projects and initiatives benefit from the scalability that partnerships provide. By pooling resources and expertise, PPPs can implement projects on a scale that individual entities might not achieve alone, resulting in greater impact and broader benefits.
- **Risk Sharing:** Partnerships allow for the distribution of risks associated with large-scale projects. Governments can offer policy support and financial incentives, businesses can share technological and operational risks, and non-profits can mitigate social and environmental risks. This risk-sharing mechanism makes ambitious projects more feasible and attractive to all parties involved.

Challenges and Considerations

- **Coordination and Governance:** Effective coordination and clear governance structures are essential for the success of PPPs. Clear roles, responsibilities, and communication channels must be established to ensure smooth collaboration and decision-making.
- **Alignment of Interests:** For partnerships to be effective, the interests and goals of all stakeholders must be aligned. This requires transparent discussions and negotiations to ensure that all parties benefit from the collaboration and are committed to the shared objectives.
- **Sustainability and Long-term Commitment:** Successful PPPs require long-term commitment from all stakeholders. Sustainability should be a core consideration in the planning and implementation phases to ensure that projects deliver lasting benefits and remains viable over time.

Conclusion

Collaboration and partnerships are vital for addressing the complex and multifaceted challenges posed by climate change. By leveraging the strengths and resources of businesses, governments, and non-profits, public-private partnerships can mobilize significant resources, share expertise, and implement impactful adaptation projects. Examples such as the Rockefeller Foundation's 100 Resilient Cities initiative and the Climate Finance Partnership illustrate the potential of PPPs to enhance resilience, drive innovation, and achieve scalable impact. Effective coordination, alignment of interests, and long-term commitment are essential for the success of these collaborations, ensuring that they deliver sustainable and lasting benefits for businesses and communities alike.

Sustainable Business Models and Practices

Adopting sustainable business models and practices is essential for aligning profit with environmental stewardship. These models prioritize long-term value creation, resource efficiency, and social responsibility.

Circular Economy

The circular economy model is a transformative approach to production and consumption that aims to minimize waste and make the most of resources. This model contrasts sharply with the traditional linear economy, which follows a 'take, make, dispose' pattern.

By adopting circular economy principles, businesses can enhance sustainability, reduce environmental impact, and create economic opportunities.

- **Designing for Longevity, Reparability, and Recyclability:** At the heart of the circular economy is the design of products that are built to last, can be easily repaired, and are made from recyclable materials. This approach reduces the need for new raw materials and minimizes waste. For instance, Patagonia, an outdoor apparel company, designs its products to be durable and offers repair services to extend the lifespan of its clothing. Patagonia's "Worn Wear" programme encourages customers to repair and reuse their gear instead of discarding it, promoting a culture of sustainability and resource efficiency.

- **Creating Closed-loop Systems:** In a circular economy, closed-loop systems ensure that materials are continuously reused and recycled, reducing the need for virgin resources and minimizing waste. IKEA, a global furniture retailer, has embraced this principle by using recycled materials in its products and offering take-back services for used furniture. IKEA's circular approach includes designing products for easy disassembly and recycling, allowing materials to be recovered and reused in new products. This not only reduces waste but also lowers the environmental footprint of IKEA's supply chain.
- **Economic and Environmental Benefits:** The circular economy offers significant economic and environmental benefits. By reducing reliance on finite resources, businesses can lower production costs and mitigate supply chain risks. Additionally, circular practices help reduce waste and pollution, contributing to environmental sustainability. The circular economy also creates new business opportunities in areas such as product refurbishment, recycling, and resource recovery, driving innovation and economic growth.

Green Supply Chains

Green supply chain management focuses on incorporating sustainable practices throughout the entire supply chain, from sourcing raw materials to manufacturing, distribution, and end-of-life disposal. By prioritizing sustainability, businesses can reduce their environmental impact, improve efficiency, and enhance their reputation among consumers.

- **Sourcing Materials Responsibly:** Responsible sourcing involves selecting materials that have minimal environmental impact and are produced ethically. Companies like Unilever have made significant commitments to sustainable sourcing.
  Unilever has pledged to source 100 per cent of its agricultural raw materials sustainably, ensuring that ingredients are produced in ways that protect ecosystems, support biodiversity, and promote fair labour practices. This commitment helps Unilever reduce its environmental footprint and ensures that its supply chain supports sustainable development goals.
- **Reducing Energy Consumption:** Energy consumption is a major contributor to greenhouse gas emissions in supply chains. By implementing energy-efficient practices and transitioning to renewable energy sources, businesses can significantly reduce their carbon footprint. For example, Apple has worked to ensure that its supply chain partners use renewable energy for manufacturing processes. Apple's Supplier Clean Energy Programme aims to have all of its supply chain partners powered by renewable energy, contributing to Apple's goal of achieving carbon neutrality.
- **Minimizing Waste:** Reducing waste throughout the production process is a key aspect of green supply chain management. This can be achieved through strategies such as lean manufacturing, waste-to-energy initiatives, and recycling programmes. Toyota, a leader in lean manufacturing, employs the "Toyota Production System" to minimize waste and improve efficiency. By continuously identifying and eliminating waste, Toyota reduces its environmental impact and enhances operational efficiency.
- **Enhancing Transparency and Accountability:** Transparency and accountability are critical components of green supply chains.

Businesses must ensure that their sustainability practices are transparent and verifiable, allowing stakeholders to assess their environmental performance. Certification schemes, such as Fair Trade and Rain Forest Alliance, provide third-party verification of sustainable practices, ensuring that products meet stringent environmental and social criteria. Companies like Nestlé and Starbucks have embraced these certifications to enhance the sustainability of their supply chains and build trust with consumers.

- **Collaboration and Innovation:** Collaboration with suppliers, customers, and other stakeholders is essential for developing and implementing green supply chain practices. Businesses can work with suppliers to improve sustainability standards, share best practices, and innovate new solutions. Collaborative platforms, such as the Sustainable Apparel Coalition, bring together brands, manufacturers, and non-profits to develop industry-wide sustainability standards and tools. By fostering collaboration and innovation, businesses can drive systemic change and improve the overall sustainability of their supply chains.

Conclusion

Adopting circular economy principles and green supply chain management are crucial strategies for sustainable business practices. The circular economy focuses on minimizing waste and maximizing resource efficiency through product design, closed-loop systems, and recycling. Companies like Patagonia and IKEA exemplify the benefits of this approach by offering repair services and using recycled materials. Green supply chains prioritize responsible sourcing, energy efficiency, waste reduction, and transparency, as demonstrated by companies like Unilever and Apple. These practices not only reduce environmental impact but also enhance operational efficiency, mitigate risks, and create new business opportunities. By integrating circular economy and green supply chain principles, businesses can align profit with planet, contributing to a sustainable and resilient future.

Corporate Social Responsibility (CSR)

Corporate Social Responsibility (CSR) initiatives are essential for businesses to demonstrate their commitment to ethical practices, community development, and environmental sustainability. By integrating CSR into their core business strategies, companies can enhance their reputation, build customer loyalty, and contribute to the well-being of society.

- **Investing in Community Development:** CSR initiatives often focus on improving the quality of life in communities where businesses operate. This can involve building infrastructure, supporting local businesses, and creating job opportunities. For example, companies like Coca Cola have invested in community development projects that provide access to clean water, sanitation, and education in developing countries. These projects not only benefit communities but also create a positive brand image for the company.
- **Supporting Education and Health Programmes:** Education and health are critical areas where businesses can make a significant impact through CSR. By supporting schools, scholarships, vocational training, and health clinics, companies can help improve educational outcomes and health standards in their communities. For instance, Johnson & Johnson runs numerous programmes aimed at improving maternal and child health, combating diseases, and providing health education. These efforts align with the company's mission to improve health outcomes globally.
- **Ensuring Fair Labour Practices:** Ethical labour practices are a fundamental aspect of CSR. Companies must ensure that their operations and supply chains uphold fair labour standards, including fair wages, safe working conditions, and the prohibition of child and forced labour. Ben & Jerry's, for example, integrates fair labour practices into its

supply chain by sourcing ingredients like cocoa and vanilla from Fair Trade certified suppliers. This commitment supports ethical labour practices and sustainable farming in developing countries.

- **Environmental Sustainability:** Businesses that prioritize CSR often focus on reducing their environmental impact. This can involve reducing greenhouse gas emissions, minimizing waste, conserving water, and using sustainable materials. Ben & Jerry's has made environmental sustainability a core part of its CSR strategy by implementing eco-friendly practices in its production processes, such as reducing carbon emissions and using sustainably sourced ingredients.
- **Social Justice:** Companies can play a significant role in promoting social justice through their CSR initiatives. This includes advocating for human rights, supporting diversity and inclusion, and addressing social inequalities. Ben & Jerry's is well-known for its commitment to social justice, actively campaigning for issues such as marriage equality, racial justice, and climate action. By taking a stand on these issues, companies can influence positive social change and build stronger connections with their customers and communities.

Energy Efficiency

Energy efficiency is a key strategy for businesses to reduce operational costs and minimize their environmental footprint. By investing in energy-efficient technologies and optimizing production processes, companies can achieve significant energy savings and contribute to sustainability goals.

- **Investing in Energy-efficient Technologies:** Businesses can reduce energy consumption by adopting energy-efficient technologies. This includes upgrading to energy-efficient lighting, heating, ventilation, and air conditioning (HVAC) systems, and using energy-efficient equipment and machinery. For example, Google and Microsoft have invested heavily in energy-efficient technologies for their data centres. These companies use advanced cooling systems, efficient servers, and artificial intelligence (AI) to optimize energy use, resulting in substantial energy savings and reduced carbon emissions.

- **Optimizing Production Processes:** Improving the efficiency of production processes is another important aspect of energy management. Companies can implement lean manufacturing techniques, automate processes, and use real-time data analytics to identify and eliminate inefficiencies. For instance, automotive manufacturers like Toyota have optimized their production processes through the Toyota Production System (TPS), which focuses on continuous improvement and waste reduction. This approach not only enhances energy efficiency but also improves overall operational efficiency.
- **Implementing Energy Management Systems:** Energy Management Systems (EMS) help businesses monitor, control, and optimize their energy use. EMS technologies provide real-time data on energy consumption, enabling companies to identify energy-saving opportunities and implement measures to reduce energy use. ISO 50001 is a globally recognized standard for energy management systems, providing a framework for businesses to develop and implement an effective EMS. Companies like Siemens and Schneider Electric offer comprehensive EMS solutions that help businesses achieve their energy efficiency goals.
- **Renewable Energy Investments:** In addition to improving energy efficiency, many companies are investing in renewable energy sources to power their operations. Google and Microsoft have committed to powering their data centres and offices with 100 per cent renewable energy. These investments in solar, wind, and other renewable energy projects not only reduce dependence on fossil fuels

but also support the transition to a sustainable energy future.

- **Economic and Environmental Benefits:** Implementing energy-efficient practices offers numerous benefits. Economically, businesses can achieve significant cost savings by reducing energy consumption and lowering utility bills. Environmentally, energy efficiency reduces greenhouse gas emissions and mitigates the impacts of climate change. Additionally, companies that prioritize energy efficiency can enhance their reputation and attract environmentally conscious customers and investors.

Conclusion

Corporate Social Responsibility (CSR) and energy efficiency are integral components of sustainable business practices. CSR initiatives allow businesses to take responsibility for their environmental and social impacts, investing in community development, supporting education and health programmes, ensuring fair labour practices, and promoting social justice.

Companies like Ben & Jerry's exemplify how integrating CSR into core business strategies can enhance brand reputation and contribute to societal well-being. Energy efficiency is another critical area where businesses can make a significant impact. By investing in energy-efficient technologies, optimizing production processes, and implementing energy management systems, companies can achieve substantial energy savings and reduce their environmental footprint. Tech giants like Google and Microsoft demonstrate the benefits of prioritizing energy efficiency and renewable energy investments. By adopting CSR and energy-efficient practices, businesses can align profit with planet, driving sustainable development and contributing to a resilient and sustainable future.

Sustainable Product Design

Sustainable product design is a cornerstone of environmentally responsible business practices. By prioritizing eco-friendly materials, energy-efficient production processes, and end-of-life considerations, companies can minimize their environmental footprint and contribute to a more sustainable future.

- **Eco-friendly Materials:** Selecting sustainable materials is a fundamental aspect of sustainable product design. Companies aim to use materials that have minimal environmental impact throughout their lifecycle, from extraction and processing to disposal. This includes renewable materials like bamboo, cork, and organic cotton, as well as recycled materials such as recycled plastics and metals. By prioritizing eco-friendly materials, companies like Tesla and Apple reduce resource depletion and environmental pollution associated with traditional manufacturing processes.
- **Energy-efficient Production:** Sustainable product design involves minimizing energy consumption during manufacturing processes. Companies strive to optimize production processes, reduce waste, and use energy-efficient machinery and equipment. Renewable energy sources, such as solar and wind power, are increasingly being used to power manufacturing facilities, further reducing carbon emissions. Tesla, for example, operates its Gigafactories using renewable energy, significantly reducing its environmental impact.
- **End-of-Life Considerations:** Considering the end-of-life disposal or recycling of products is essential for sustainable product design. Companies design products with disassembly and recyclability in mind, ensuring that materials can be easily separated and recycled at the end of their useful life. Biodegradable materials are also used where possible to minimize environmental pollution. Apple, for instance, incorporates recyclable materials into its products and offers recycling programmes to customers to responsibly dispose of old devices.

- **Life Cycle Assessment (LCA):** Conducting Life Cycle Assessments (LCAs) is a valuable tool for evaluating the environmental impact of products throughout their lifecycle. LCAs consider factors such as raw material extraction, manufacturing, transportation, product use, and end-of-life disposal to identify opportunities for improvement. By conducting LCAs, companies can identify areas where they can reduce environmental impact and make informed decisions about sustainable product design.
- **Certifications and Standards:** Many companies seek certifications and adhere to environmental standards to demonstrate their commitment to sustainable product design. Certifications such as Cradle to Cradle (C2C) and Forest Stewardship Council (FSC) ensure that products meet rigorous environmental and social criteria. Adhering to these standards provides credibility and transparency to consumers, who increasingly value environmentally friendly products. For example, Tesla's electric vehicles are certified as environmentally sustainable and meet stringent emissions standards.
- **Consumer Education and Awareness:** Educating consumers about the environmental benefits of sustainable products are essential for driving demand and promoting sustainable consumption behaviours. Companies engage in marketing and educational campaigns to raise awareness about the environmental impact of their products and the benefits of choosing sustainable alternatives. By empowering consumers to make informed choices, companies like Tesla and Apple promote a culture of sustainability and environmental stewardship.

Conclusion

In conclusion, sustainable product design plays a crucial role in mitigating environmental impact and promoting a more sustainable future. By prioritizing eco-friendly materials, energy-efficient production processes, and end-of-life considerations, companies can minimize their environmental footprint and contribute to a circular economy. Through certifications, consumer education, and transparent communication, companies can build trust with consumers and drive demand for sustainable products. Sustainable product design is not only environmentally responsible but also economically viable, as companies increasingly recognize the value of aligning profit with planet.

Economic Benefits of Sustainability

Sustainable business practices offer numerous economic benefits, enhancing profitability, resilience, and competitiveness.

Cost Savings

One of the most compelling arguments for adopting sustainable business practices is the potential for significant cost savings. Companies that prioritize energy efficiency, waste reduction, and resource optimization can reap financial benefits while also reducing their environmental footprint.

- **Energy Efficiency:** Implementing energy-efficient measures not only reduces greenhouse gas emissions but also lowers utility bills and operational expenses for businesses. By investing in energy-efficient technologies, upgrading lighting systems, and optimizing heating, ventilation, and air conditioning (HVAC) systems, companies can substantially reduce their energy consumption and associated costs over time. For example, switching to LED lighting can lead to considerable energy savings, as LEDs are more energy-efficient and have a longer lifespan than traditional incandescent bulbs.
- **Waste Reduction:** Reducing waste generation and improving waste management practices can also result in significant cost savings for businesses. By implementing recycling programmes, reusing materials, and minimizing

packaging waste, companies can reduce disposal costs and potentially generate revenue from recycled materials. Additionally, reducing waste can lead to efficiencies in production processes, such as streamlined workflows and reduced material losses, further contributing to cost savings.

- **Resource Optimization:** Sustainable business practices often involve optimizing resource use, such as water, raw materials, and natural resources. By implementing measures to conserve water, minimize raw material usage, and optimize supply chain processes, companies can reduce their dependence on finite resources and associated costs. For example, water-efficient technologies in manufacturing processes can lead to substantial savings in water consumption and wastewater treatment costs.

Risk Mitigation

In addition to cost savings, adopting sustainable business practices can help companies mitigate various risks, including those related to resource scarcity, regulatory changes, and reputational damage.

- **Resource Scarcity:** As global demand for resources continues to rise, companies face increasing risks associated with resource scarcity and supply chain disruptions. By adopting sustainable practices that prioritize resource efficiency and diversification, businesses can reduce their vulnerability to shortages and price fluctuations. For example, companies that invest in renewable energy sources are less susceptible to volatility in fossil fuel prices and regulatory changes affecting carbon emissions.
- **Regulatory Changes:** Regulatory frameworks around the world are evolving to address environmental and social challenges, including climate change, pollution, and labour rights. Companies that proactively align with emerging regulations and industry standards can avoid compliance issues and associated penalties. By integrating sustainability

into their business strategies and operations, companies can stay ahead of regulatory changes and maintain a competitive advantage in the market.

- **Reputational Damage:** Public perception of a company's environmental and social responsibility can significantly impact its brand reputation and customer loyalty. Businesses that prioritize sustainability and corporate social responsibility (CSR) are better positioned to build trust with consumers, employees, investors, and other stakeholders. By demonstrating a commitment to ethical business practices, environmental stewardship, and social impact, companies can enhance their brand reputation and mitigate the risk of reputational damage from negative publicity or controversies.

Conclusion

In conclusion, the adoption of sustainable business practices offers not only cost savings but also risk mitigation benefits for companies. By prioritizing energy efficiency, waste reduction, and resource optimization, businesses can reduce operational expenses and enhance their resilience to environmental, regulatory, and reputational risks.

As sustainability becomes increasingly important in the business world, companies that embrace sustainable practices are better positioned to thrive in a rapidly changing and competitive global marketplace.

Market Opportunities

Embracing sustainability not only reduces costs and mitigates risks but also unlocks new market opportunities for businesses. By aligning their products and services with environmental and social values, companies can tap into growing consumer demand for sustainable alternatives and gain a competitive advantage in the marketplace.

- **Consumer Demand:** Increasingly, consumers are prioritizing sustainability when making purchasing decisions. They seek products and services that minimize environmental impact, promote social responsibility, and support ethical business practices. Companies that offer sustainable options can capture a growing market segment of environmentally conscious consumers, driving sales and brand loyalty. For example, consumers are increasingly choosing eco-friendly alternatives in sectors such as fashion, food, and personal care, creating opportunities for businesses to innovate and differentiate themselves.
- **Green Markets:** The rise of sustainability-conscious consumers has spurred the growth of green markets, where environmentally friendly products and services are in high demand. These markets encompass a wide range of industries, including renewable energy, organic agriculture, eco-tourism, and green technology. Businesses that invest in sustainable practices can access these lucrative markets and capitalize on the growing demand for eco-friendly solutions. For instance, companies specializing in renewable energy technologies, such as solar panels and wind turbines, are well-positioned to thrive in the transition to a low-carbon economy.
- **Competitive Advantage:** Embracing sustainability can provide businesses with a competitive edge in crowded marketplaces. By differentiating themselves as environmentally responsible brands, companies can attract environmentally conscious consumers who are willing to pay a premium for sustainable products and services. Moreover, sustainability initiatives can enhance brand reputation, build trust with customers, and foster long-term relationships. For example, companies that demonstrate a commitment to environmental stewardship and social responsibility often enjoy higher levels of customer loyalty and positive brand perception.

Investment and Financing

In addition to capturing new market opportunities, sustainable businesses benefit from access to specialized investment and financing options that prioritize environmental, social, and governance (ESG) criteria. Investors and lenders increasingly recognize the value of sustainability and seek out companies that demonstrate strong ESG performance.

- **Green Financing:** Green financing encompasses a range of financial products and instruments designed to support sustainable projects and initiatives. These include climate bonds, green loans, and sustainability-linked bonds, which provide capital for investments in renewable energy, energy efficiency, sustainable infrastructure, and other environmentally beneficial projects. Businesses that prioritize sustainability can access green financing options to fund their initiatives and accelerate their transition to a low-carbon economy. For example, companies may use proceeds from green bonds to finance the installation of solar panels, upgrade to energy-efficient equipment, or implement sustainable supply chain practices.
- **ESG Investing:** Environmental, social, and governance (ESG) investing integrates sustainability considerations into investment decision-making. ESG investors evaluate companies based on their performance in areas such as carbon emissions, resource efficiency, labour practices, and corporate governance. Businesses that excel in ESG metrics are more attractive to investors seeking to align their portfolios with sustainability goals. By demonstrating strong ESG performance, companies can attract investment capital, reduce financing costs, and enhance shareholder value.

  For instance, companies that prioritize transparency, accountability, and ethical conduct are more likely to attract socially responsible investors who value sustainability and corporate responsibility. In summary, sustainability offers not only market opportunities but also access to specialized investment and financing options for businesses. By aligning their strategies with environmental and social values, companies can differentiate themselves, attract environmentally conscious consumers, and access capital to fund sustainable projects and initiatives. As sustainability continues to gain momentum in the business world, companies that embrace sustainable practices are well-positioned to thrive in a changing and increasingly competitive market landscape.

- **Brand Loyalty and Reputation:** Companies committed to sustainability can build stronger brand loyalty and enhance their reputation. Consumers and stakeholders are increasingly valuing businesses that demonstrate environmental and social responsibility. This positive perception can lead to increased customer retention, employee satisfaction, and stakeholder trust.

Case Studies of Successful Sustainable Businesses

- **Patagonia:** Patagonia is a leading example of a business that prioritizes sustainability and social responsibility. The company is committed to using sustainable materials, reducing environmental impacts, and supporting environmental activism. Patagonia's "Worn Wear" programme encourages customers to repair, reuse, and recycle their clothing, promoting a circular economy. The company's dedication to sustainability has not only enhanced its brand reputation but also driven customer loyalty and profitability.
- **Unilever:** Unilever has integrated sustainability into its core business strategy through its Sustainable Living Plan. The Company aims to decouple its growth from environmental impact while increasing its positive social impact. Unilever has made significant progress in reducing greenhouse

gas emissions, improving water efficiency, and sourcing sustainable raw materials. The company's commitment to sustainability has resulted in cost savings, risk reduction, and enhanced brand value.

- **Tesla:** Tesla's mission to accelerate the world's transition to sustainable energy has positioned it as a leader in the clean energy and electric vehicle industries. The company's innovative products, such as electric cars, solar panels, and energy storage solutions, contribute to reducing carbon emissions and promoting renewable energy. Tesla's focus on sustainable product design and manufacturing has driven significant market growth and investor interest.
- **IKEA:** IKEA is committed to becoming a circular and climate-positive business by 2030. The company focuses on sustainable sourcing, energy efficiency, and waste reduction. IKEA's "People & Planet Positive" strategy includes initiatives to use renewable and recycled materials, improve energy efficiency in stores and production, and promote sustainable living among customers. These efforts have enhanced IKEA's sustainability performance and competitive advantage.

Conclusion

Sustainable business practices are essential for aligning profit with environmental stewardship and social responsibility. Businesses play a crucial role in climate adaptation by adopting sustainable models, investing in green technologies, and collaborating with stakeholders. The economic benefits of sustainability, including cost savings, risk mitigation, market opportunities, and enhanced reputation, demonstrate that sustainable practices are not only good for the planet but also good for business. Case studies of companies like Patagonia, Unilever, Tesla, and IKEA illustrate the successful implementation of sustainable practices and their positive impact on profitability and resilience. By prioritizing sustainability, businesses can contribute to a more sustainable and resilient future while achieving long-term success.

REFERENCES

Aghion, P., Bergeaud, A., Cette, G. and Blundell, R. (2022). "The Impact of Inflation on Innovation: Evidence from Patent Data". *Journal of Monetary Economics, 131*: 77-92.

Bernanke, B.S. (2022). "The Real Effects of Disinflation". *Brookings Papers on Economic Activity, 2022*(1): 77-123.

Blinder, A.S. (2022). "Macroeconomics Under the Influence: Inflation, Expectations, and Policy". *Journal of Economic Perspectives, 36*(4): 65-88.

Bordo, M.D. and Levin, A.T. (2022). "Central Bank, Digital Currency and the Future of Monetary Policy". *National Bureau of Economic Research Working Paper*, No. 29923.

Cogley, T. and Sbordone, A.M. (2022). "Inflation and the Evolution of Price Dispersion". *Journal of Monetary Economics, 122*: 156-170.

Daly, M.C. and Hobijn, B. (2022). "Wage Growth and Inflation: The Role of Productivity and Labour Market Slack". *Economic Review, Federal Reserve Bank of Kansas City, 2022*(Q2): 47-74.

Draghi, M. (2022). "Policy Coordination in a High-Inflation Environment". *Journal of International Economics*, Vol. 132, Editorial Board Article No. 103594.

Drechsler, I., Savov, A. and Schnabl, P. (2022). "The Financial Stability Implications of Inflation". *Journal of Finance, 77*(4): 1873-1911.

Eggertsson, G.B. and Woodford, M. (2022). "The Zero Bound on Interest Rates and Optimal Monetary Policy". *Brookings Papers on Economic Activity, 2022*(2): 139-211.

Gertler, M. and Karadi, P. (2022). "Monetary Policy Surprises, Credit Costs, and Economic Activity". *American Economic Journal: Macroeconomics, 14*(2): 78-117.

Goodhart, C.A.E. and Pradhan, M. (2022). "The Great Demographic Reversal and Inflation". *Journal of Economic Dynamics and Control*, Vol. 135, Editorial Board Article No. 104297.

Gourinchas, P.O. (2022). "Global Economic Challenges: Inflation, Trade, and Policy". *IMF Economic Review, 70*(3): 249-278.

Guvenen, F., Kaplan, G. and Song, J. (2022). "How Does Inflation Affect Different Income Groups?" *Journal of Political Economy, 130*(5): 1437-1473.

Hamilton, J.D. (2022). "Why You Should Never Use the Hodrick-Prescott Filter". *Review of Economic Studies, 89*(1): 104-120.

Jordà, Ò. and Taylor, A.M. (2022). "The Pandemic, Inflation, and Debt Sustainability". *Economic Policy, 37*(110): 275-314.

Kiley, M.T. (2022). "Unemployment, Inflation, and Monetary Policy: Evidence from Advanced and Emerging Market Economies". *Journal of International Money and Finance*, Vol. 122, Editorial Board Article No. 102564.

Laubach, T. and Williams, J.C. (2022). "Measuring the Natural Rate of Interest: International Trends and Determinants". *Brookings Papers on Economic Activity, 2022*(3): 57-108.

Mishkin, F.S. (2022). "Monetary Policy Strategy: Lessons from the Financial Crisis". *Journal of Economic Perspectives, 36*(3): 123-148.

Mussa, M. (2022). "Inflation, Real Interest Rates, and the Real Exchange Rate: An Empirical Investigation". *Journal of Monetary Economics, 127*: 243-271.

Orphanides, A. (2022). "Policy Rules and Economic Stability: Lessons from the Great Inflation". *Journal of Economic Dynamics and Control*, Vol. 135, Editorial Board Article No. 104313.

Reis, R. (2022). "The Analytics of Monetary Policy: A New Keynesian Perspective". *Quarterly Journal of Economics, 137*(4): 1951-1987.

Sims, C.A. (2022). "Fiscal Policy, Inflation, and Debt Sustainability". *Journal of Economic Theory*, Vol. 198, Editorial Board Article No. 105297.

Smets, F. and Wouters, R. (2022). "An Estimated Dynamic Stochastic General Equilibrium Model of the Euro Area". *Journal of Economic Dynamics and Control*, Vol. 133, Editorial Board Article No. 104295.

Stock, J.H. and Watson, M.W. (2022). "Disentangling the Channels of the 2021-2022 Inflation Surge". *Journal of Applied Econometrics, 37*(6): 975-996.

11

Social Justice and Economic Adaptation

Addressing Inequities

Intersection of Social Justice and Economic Adaptation

The intersection of social justice and economic adaptation to climate change represents a critical aspect of addressing the multifaceted challenges posed by environmental degradation and its disproportionate impacts on marginalized communities. At its core, social justice encompasses principles of fairness, equity, and inclusivity, aiming to ensure that all individuals and communities have equal access to resources, opportunities, and protections. Economic adaptation, on the other hand, involves implementing strategies to mitigate the adverse effects of climate change on economies, infrastructure, and livelihoods. Understanding and addressing the intersection of these two concepts is essential for fostering resilience and promoting sustainable development in the face of environmental challenges.

One of the key aspects of the intersection between social justice and economic adaptation is the recognition of the differential vulnerabilities and exposures faced by marginalized communities. Historically marginalized groups, including low-income households, communities of colour, indigenous peoples, and individuals with disabilities, often experience disproportionate impacts from climate change and environmental degradation. These disparities are rooted in systemic inequalities, including unequal access to resources, discriminatory policies, and patterns of environmental injustice. For example, marginalized communities may be more likely to live in areas prone to flooding, pollution, or extreme heat due to factors such as residential segregation and lack of political power.

Addressing these vulnerabilities requires a comprehensive approach that goes beyond traditional economic adaptation strategies. It involves recognizing the root causes of social and environmental injustices and working to dismantle systems of oppression that perpetuate inequality. This includes engaging directly with affected communities to understand their needs, priorities, and experiences, and incorporating their perspectives into decision-making processes. By centring the voices of marginalized groups in adaptation efforts, policymakers and stakeholders can develop more effective and equitable strategies that address the specific challenges faced by different communities. Inclusive policies and practices are essential for ensuring that economic adaptation efforts benefit all members of society equitably.

This involves designing adaptation measures that prioritize the needs of marginalized communities and promote social cohesion and solidarity. For example, initiatives such as community-led climate resilience planning and participatory budgeting can empower local residents to identify adaptation priorities and allocate resources in ways that address their unique circumstances. In addition, policies that promote affordable housing, equitable access to healthcare, and quality education can help build the resilience of vulnerable communities and reduce their exposure to climate-related risks. Furthermore, economic adaptation strategies must consider the intersectionality of social identities and experiences, recognizing that individuals and communities may face multiple forms of oppression and discrimination simultaneously.

For example, women, particularly those from marginalized backgrounds, often bear a disproportionate burden of climate impacts due to gender inequalities in access to resources, decision-making power, and social status. Similarly, LGBTQ+ individuals may face unique challenges related to discrimination, stigma, and exclusion in disaster response and recovery efforts. By adopting an intersectional approach to economic adaptation, policymakers can better understand the complex ways in which social identities intersect with environmental vulnerabilities and tailor interventions accordingly. Case studies provide valuable insights into equitable adaptation strategies that prioritize social justice and inclusivity. For example, the Resilience Justice Project in New Orleans, Louisiana, aims to address the root causes of vulnerability and inequality in the face of climate change.

The project brings together community organizers, researchers, and policymakers to develop adaptation strategies that prioritize the needs of historically marginalized communities, including low-income residents, people of colour, and indigenous groups. Through a participatory approach, the project seeks to build community resilience by addressing issues such as affordable housing, access to healthcare, and environmental justice. Similarly, the Navajo Nation in the south-western United States has implemented a range of adaptation measures to address the impacts of climate change on indigenous communities. These include investments in renewable energy, water conservation, and sustainable agriculture, as well as efforts to protect cultural heritage and traditional knowledge. By integrating indigenous perspectives and practices into adaptation planning and decision-making processes, the Navajo Nation aims to build resilience while preserving its cultural identity and sovereignty.

Conclusion

In conclusion, the intersection of social justice and economic adaptation represents a crucial aspect of addressing the challenges of climate change and environmental degradation. By recognizing and addressing the differential vulnerabilities and exposures faced by marginalized communities, adopting inclusive policies and practices, and embracing an intersectional approach to adaptation, policymakers and stakeholders can develop more effective and equitable strategies that promote resilience and sustainability for all. Case studies provide valuable examples of how these principles can be put into practice, highlighting the importance of centring the voices and experiences of affected communities in adaptation efforts.

Addressing Vulnerabilities and Inequalities

Addressing vulnerabilities and inequalities in the context of climate change adaptation is essential for promoting social justice and building resilient communities. Vulnerabilities and inequalities arise from a complex interplay of social, economic, and environmental factors, which can exacerbate the impacts of climate change on marginalized populations. To effectively address these challenges, it is crucial to understand the root causes of vulnerability and inequality and implement targeted interventions that prioritize the needs of the most marginalized groups. One of the key factors contributing to vulnerability and inequality is unequal access to resources and opportunities. Historically marginalized communities, including low-income households, communities of colour, indigenous peoples, and individuals with disabilities, often face systemic barriers that limit their access to essential resources such as clean water, healthcare, and education.

These disparities can increase their susceptibility to the adverse effects of climate change, such as extreme weather events, sea-level rise, and food insecurity. Environmental injustices, including the disproportionate siting of hazardous waste facilities, pollution, and lack of access to green spaces, further exacerbate vulnerabilities and inequalities. For example, marginalized communities are more likely to live in areas with poor air quality, which can exacerbate respiratory illnesses and other health problems. Additionally, communities located in flood-prone or environmentally degraded areas may lack the resources to relocate to safer locations, trapping them in cycles of poverty and vulnerability. Addressing vulnerabilities and inequalities requires a multi-dimensional approach that tackles the root causes of social and environmental injustices. This includes implementing policies and programmes that promote equitable access to resources, opportunities, and protections.

For example, investing in affordable housing, public transportation, and green infrastructure can improve the quality of life for marginalized communities while reducing their exposure to climate-related risks. Similarly, expanding access to healthcare, education, and social services can enhance the resilience of vulnerable populations and empower them to adapt to changing environmental conditions. Community-driven approaches to climate adaptation are essential for addressing vulnerabilities and inequalities effectively. Engaging directly with affected communities to understand their needs, priorities, and experiences is crucial for developing targeted interventions that reflect their unique circumstances. Community-based organizations, grassroots movements, and indigenous leaders can play a key role in driving adaptation efforts and advocating for policies that promote social justice and equity.

Inclusive decision-making processes that centre the voices of marginalized communities are essential for ensuring that adaptation strategies are responsive to their needs and priorities. This involves creating opportunities for meaningful participation and collaboration among diverse stakeholders, including community members, policymakers, researchers, and practitioners. By fostering partnerships and dialogue, stakeholders can develop more inclusive and effective adaptation measures that address the root causes of vulnerability and inequality. Furthermore, addressing vulnerabilities and inequalities requires a commitment to social justice and human rights principles. This includes recognizing and addressing historical injustices, such as colonialism, racism, and discrimination that have contributed to the marginalization of certain groups. Reconciliation processes, reparative justice initiatives, and land rights recognition are essential steps towards building trust and promoting equity in climate adaptation efforts.

Case studies provide valuable insights into successful approaches for addressing vulnerabilities and inequalities in the context of climate change adaptation. For example, the Green Climate Fund's Readiness and Preparatory Support Programme supports developing countries in enhancing their capacity to access climate finance and implement adaptation projects. Through targeted capacity-building activities and technical assistance, the programme aims to strengthen the resilience of vulnerable communities and promote inclusive and equitable adaptation. Similarly, the Participatory Urban Planning project in Salvador, Brazil, engages local residents, civil society organizations, and government agencies in collaborative decision-making processes to address climate-related risks and vulnerabilities in informal settlements.

By prioritizing the needs and perspectives of marginalized communities, the project aims to build resilience and promote social inclusion in urban adaptation planning.

Conclusion

In conclusion, addressing vulnerabilities and inequalities is essential for building resilience and promoting social justice in the face of climate change. By understanding the root causes of vulnerability and inequality, implementing targeted interventions that prioritize the needs of marginalized communities, and fostering inclusive decision-making processes, policymakers and stakeholders can develop more effective and equitable adaptation strategies. Case studies provide valuable examples of successful approaches for addressing vulnerabilities and inequalities, highlighting the importance of community-driven, participatory, and rights-based approaches to climate adaptation.

Inclusive Policies and Practices

Inclusive policies and practices are essential for promoting social justice and ensuring that climate adaptation efforts benefit all members of society, particularly those who are most vulnerable and marginalized. Inclusivity involves recognizing and valuing the diversity of experiences, needs, and perspectives within communities and integrating them into decision-making processes, policy development, and implementation strategies. By prioritizing equity, participation, and representation, inclusive policies and practices can help address systemic inequalities and build resilience to climate change in a way that benefits everyone. At the heart of inclusive policies and practices is the principle of equity, which involves recognizing and addressing historical and systemic injustices that have marginalized certain groups.

Equity acknowledges that different communities have different needs and vulnerabilities based on factors such as race, ethnicity, gender, socio-economic status, age, disability, and geographic location. Inclusive policies seek to redress these disparities by ensuring that resources, opportunities, and protections are distributed fairly and that no group is left behind. One key aspect of inclusive policies and practices is meaningful participation, which involves engaging diverse stakeholders in decision-making processes and empowering them to contribute to shaping policies and programmes that affect their lives. Meaningful participation goes beyond tokenistic involvement and requires creating opportunities for marginalized communities to have a genuine voice in decision-making, from the identification of priorities and the development of solutions to the implementation and evaluation of initiatives.

Meaningful participation can take various forms, including community consultations, participatory planning processes, citizen assemblies, and collaborative governance structures. These approaches allow communities to articulate their needs, express their concerns, and contribute their knowledge and expertise to the design and implementation of climate adaptation measures. By actively involving affected communities in decision-making, policymakers can ensure that adaptation efforts are responsive to local contexts, culturally appropriate, and effective in addressing the root causes of vulnerability and inequality. Inclusive policies and practices also prioritize representation, ensuring that the voices of marginalized groups are heard and their perspectives are considered in decision-making processes. This involves actively recruiting and supporting diverse leaders and representatives from underrepresented communities to participate in governance bodies, advisory committees, and decision-making forums.

Representation helps ensure that the interests and concerns of marginalized groups are adequately represented in policy discussions and that decision-makers are accountable to those they serve. Moreover, inclusive policies and practices recognize the importance of intersectionality, which acknowledges that individuals and communities may experience multiple forms of oppression and discrimination simultaneously. Intersectional approaches to climate adaptation take into account the intersecting identities and experiences of individuals, recognizing that vulnerability and resilience are shaped by complex and interconnected factors. By addressing the intersecting dimensions of inequality, such as race, class, gender, and disability, inclusive policies can better meet the diverse needs of marginalized communities and promote social justice.

Inclusive policies and practices are also characterized by a commitment to accessibility and inclusivity, ensuring that all members of society have equal access to information, resources, and opportunities to participate in decision-making processes. This involves removing barriers to participation, such as language barriers, physical barriers, and financial barriers, and providing accommodations and support services to ensure that everyone can engage meaningfully in climate adaptation efforts. Furthermore, inclusive policies and practices prioritize community-driven approaches to climate adaptation, recognizing that communities are best positioned to identify their own needs, assets, and priorities and develop solutions that are tailored to their specific contexts. Community-driven approaches empower local residents to take ownership of adaptation initiatives and build resilience from the ground up, fostering a sense of ownership, agency, and collective action among community members.

Inclusive policies and practices are also characterized by a commitment to transparency, accountability, and feedback mechanisms, ensuring that decision-making processes are open, responsive, and accountable to the communities they serve. Transparency involves providing clear and accessible information about decision-making processes, policies, and outcomes, while accountability involves holding decision-makers and institutions responsible for their actions and decisions. Feedback mechanisms allow communities to provide input, raise concerns, and offer feedback on climate adaptation efforts, enabling policymakers to make informed decisions and adjust strategies based on community input.

By fostering transparency, accountability, and feedback, inclusive policies and practices can build trust, legitimacy, and public support for climate adaptation efforts, strengthening social cohesion and resilience in the face of climate change.

Conclusion

In conclusion, inclusive policies and practices are essential for promoting social justice, equity, and resilience in the face of climate change. By prioritizing meaningful participation, representation, accessibility, and community-driven approaches, policymakers can ensure that climate adaptation efforts benefit all members of society, particularly those who are most vulnerable and marginalized. Inclusive policies and practices foster collaboration, empowerment, and solidarity, enabling communities to build resilience and thrive in a changing climate.

Case Studies on Equitable Adaptation Strategies

Case studies offer valuable insights into how equitable adaptation strategies can be implemented in diverse contexts to address vulnerabilities and promote social justice in the face of climate change. By examining successful initiatives from around the world, we can learn from best practices, identify key challenges, and develop effective approaches to building resilience in vulnerable communities. Here, we explore several case studies that highlight innovative and equitable adaptation strategies in different regions and sectors.

Community-based Adaptation in Bangladesh

Bangladesh is one of the most climate-vulnerable countries in the world, facing threats from sea-level rise, cyclones, and floods. In response to these challenges, community-based adaptation (CBA) initiatives have been implemented to empower local communities and build resilience from the ground up. One example is the Coastal Livelihood and Environmental Action Network (CLEAN), which works with coastal communities to implement adaptation measures such as building embankments, promoting saline-tolerant crops, and providing alternative livelihood options.

By involving local residents in decision-making and implementation processes, CBA initiatives in Bangladesh have helped communities adapt to climate change while addressing underlying social and economic inequalities.

Indigenous-led Adaptation in Canada

Indigenous communities in Canada are disproportionately affected by climate change due to their reliance on natural resources and traditional livelihoods. In response, indigenous-led adaptation initiatives have emerged to integrate traditional knowledge and practices with modern climate science and technology. For example, the Inuit of Nunavut have developed a community-based monitoring programme to track environmental changes and inform adaptation strategies. Similarly, the Mi'kmaq of Atlantic Canada are revitalizing traditional fisheries and forest management practices to enhance resilience and food security. By centring indigenous perspectives and leadership in adaptation efforts, these initiatives promote cultural continuity, self-determination, and resilience in indigenous communities.

Gender-Responsive Adaptation in Nepal

Women are often disproportionately affected by climate change due to their roles as caregivers, food providers, and water managers. In Nepal, gender-responsive adaptation initiatives have been implemented to address the unique needs and vulnerabilities of women in rural communities. One example is the Rural Women's Climate Resilience Project, which trains women in climate-smart agriculture, water management, and disaster preparedness. These initiatives not only enhance women's adaptive capacity but also promote gender equality, economic empowerment, and social inclusion in rural Nepal. By recognizing and addressing gender disparities in adaptation planning and implementation, these initiatives contribute to more equitable and effective climate resilience strategies.

Urban Heat Resilience in the United States

Heatwaves are becoming more frequent and intense due to climate change, posing significant risks to urban populations, especially in low-income communities and communities of colour. In response, cities like Los Angeles have implemented heat resilience initiatives to protect vulnerable residents and reduce heat-related health risks. One example is the Cool Streets programme, which uses reflective pavement and tree planting to reduce urban heat islands in disadvantaged neighbourhoods. Similarly, the Heat Relief Network provides cooling centres, shade structures, and outreach services to vulnerable populations during heat waves.

By targeting resources and interventions to the most vulnerable communities, these initiatives promote equity and social justice in urban climate adaptation.

Ecosystem-based Adaptation in Costa Rica

Ecosystem-based Adaptation (EbA) approaches leverage the natural environment to build resilience and reduce climate risks. In Costa Rica, EbA initiatives have been implemented to protect ecosystems such as forests, wetlands, and coral reefs, which provide essential services such as flood protection, water regulation, and biodiversity conservation. For example, the Payment for Ecosystem Services (PES) programme incentivizes landowners to conserve forests and restore degraded lands, thereby enhancing resilience to climate change and supporting local livelihoods. By integrating EbA into national policies and programmes, Costa Rica has demonstrated how nature-based solutions can promote equity, sustainability, and resilience in the face of climate change.

These case studies illustrate the diverse ways in which equitable adaptation strategies can be implemented to address vulnerabilities and promote social justice in the face of climate change. From community-based approaches in Bangladesh to indigenous-led initiatives in Canada, gender-responsive programmes in Nepal, urban heat resilience in the United States, and ecosystem-based solutions in Costa Rica, these examples highlight the importance of centring community needs, promoting participation and representation, and addressing underlying social and economic inequalities in adaptation planning and implementation. By learning from these experiences and scaling up successful approaches, we can build more resilient and equitable societies in a changing climate.

REFERENCES

Auerbach, A.J. and Gorodnichenko, Y. (2012). "Fiscal Multipliers in Recessions and Expansions: Does It Matter Whether Government Spending is Increasing or Decreasing?" *Journal of Economic Perspectives*, *36*(1): 41-64.

Eichenbaum, M.S., Rebelo, S. and Wong, A. (2022). "State-Dependent Effects of Monetary Policy: The Asymmetric Role of Supply Constraints". *Review of Economic Studies*, *89*(4): 721-761.

Gourinchas, P.O. (2022). "Global Economic Challenges: Inflation, Trade, and Policy". *IMF Economic Review*, *70*(3): 249-278.

Guerrieri, V., Lorenzoni, G., Straub, L. and Werning, I. (2022). "Macroeconomic Implications of COVID-19: Can Negative Supply Shocks Cause Demand Shortages?" *American Economic Review: Insights*, *4*(2): 173-190.

Guvenen, F., Kaplan, G. and Song, J. (2022). "How Does Inflation Affect Different Income Groups?" *Journal of Political Economy, 130*(5): 1437-1473.

Hamilton, J.D. (2022). "Why You Should Never Use the Hodrick-Prescott Filter". *Review of Economic Studies, 89*(1): 104-120.

Jordà, Ò. and Taylor, A.M. (2022). "The Pandemic, Inflation, and Debt Sustainability". *Economic Policy, 37*(110): 275-314.

Kiley, M.T. (2022). "Unemployment, Inflation, and Monetary Policy: Evidence from Advanced and Emerging Market Economies". *Journal of International Money and Finance*, Vol. 122, Editorial Board Article No. 102564.

Krugman, P. (2022). "Inflation and the Power of Economic Ideas". *Foreign Affairs, 101*(2): 23-37.

Lane, P.R. (2022). "Global Supply Chains and Inflation: The Role of the Pandemic and Policy Responses". *IMF Economic Review, 70*(2): 121-143.

Laubach, T. and Williams, J.C. (2022). "Measuring the Natural Rate of Interest: International Trends and Determinants". *Brookings Papers on Economic Activity, 2022*(3): 57-108.

Leduc, S. and Wilson, D. (2022). "What is Driving the Rise in Inflation? A Primer". *Economic Letter*, Federal Reserve Bank of San Francisco, pp. 2022-27.

Martin, F. (2022). "Inflation, Wage Setting and the Monetary Transmission Mechanism". *Journal of Economic Theory*, Vol. 197, Editorial Board Article No. 105299.

Mian, A., Sufi, A. and Verner, E. (2022). "Household Debt and Inflation: The Role of Mortgage Debt in Monetary Policy Transmission". *Journal of Financial Economics, 143*(3): 1041-1061.

Mishkin, F.S. (2022). "Monetary Policy Strategy: Lessons from the Financial Crisis". *Journal of Economic Perspectives, 36*(3): 123-148.

Mussa, M. (2022). "Inflation, Real Interest Rates, and the Real Exchange Rate: An Empirical Investigation". *Journal of Monetary Economics, 127*: 243-271.

Orphanides, A. (2022). "Policy Rules and Economic Stability: Lessons from the Great Inflation". *Journal of Economic Dynamics and Control*, Vol. 135, Editorial Board Article No. 104313.

Reis, R. (2022). "The Analytics of Monetary Policy: A New Keynesian Perspective". *Quarterly Journal of Economics, 137*(4): 1951-1987.

Sims, C.A. (2022). "Fiscal Policy, Inflation, and Debt Sustainability". *Journal of Economic Theory*, Vol. 198, Editorial Board Article No. 105297.

Smets, F. and Wouters, R. (2022). "An Estimated Dynamic Stochastic General Equilibrium Model of the Euro Area". *Journal of Economic Dynamics and Control*, Vol. 133, Editorial Board Article No. 104295.

Stock, J.H. and Watson, M.W. (2022). "Disentangling the Channels of the 2021-2022 Inflation Surge". *Journal of Applied Econometrics*, *37*(6): 975-996.

Summers, L.H. and Fatás, A. (2022). "The Permanent Effects of Fiscal Consolidations". *Journal of Monetary Economics*, *124*: 144-162.

12

Future Directions for Economic Adaptation to Climate Change

As the impacts of climate change continue to manifest, it is imperative to explore future directions for economic adaptation to mitigate risks, promote sustainability, and foster global cooperation. This chapter examines emerging trends, innovations, and strategies for long-term economic planning in the face of climate uncertainty, emphasizing the importance of collaboration and knowledge sharing to address this global challenge.

Emerging Trends and Innovations in Economic Adaptation

Emerging trends in economic adaptation to climate change are driven by the urgent need to enhance resilience and sustainability in the face of environmental challenges. Businesses, governments, and communities are increasingly embracing innovative approaches and technologies to mitigate risks and seize opportunities in a changing climate. One significant trend is the rise of green finance, which involves investments in environmentally sustainable projects and initiatives. Green bonds, for example, have emerged as a popular financial instrument for funding climate-resilient infrastructure and renewable energy projects. Investors and policymakers alike are recognizing the importance of integrating climate risk into financial decision-making processes, driving demand for green investments and incentivizing businesses to prioritize sustainability in their operations.

Alongside green finance, sustainable infrastructure is another key trend in economic adaptation. As the impacts of climate change intensify, there is growing recognition of the need to build infrastructure that can withstand extreme weather events and reduce greenhouse gas emissions. Climate-resilient infrastructure projects, such as seawalls, flood barriers, and green roofs, are gaining traction as cities and regions seek to enhance their resilience to climate-related risks. These projects not only provide protection against the impacts of climate change but also create jobs, stimulate economic growth, and improve the quality of life for communities. In addition to financial and infrastructure solutions, digital technology is playing a crucial role in advancing economic adaptation to climate change. Advancements in artificial intelligence, remote sensing, and data analytics are revolutionizing climate modelling, disaster preparedness, and early warning systems. For example, sophisticated climate models can simulate future climate scenarios and assess the potential impacts on various sectors, helping businesses and governments make informed decisions about adaptation strategies.

Remote sensing technologies, such as satellite imagery and drones, are enabling more accurate monitoring of environmental changes, allowing for timely responses to natural disasters and extreme weather events. Furthermore, digital platforms and mobile applications are empowering communities to access real-time information and resources to mitigate risks and build resilience. Nature-based solutions are also emerging as a promising approach to economic adaptation to climate change. These solutions leverage the power of ecosystems to provide natural defences against climate-related hazards, such as flooding, erosion, and heat waves. Examples include restoring wetlands, mangroves, and forests to absorb carbon dioxide, regulate water flow, and protect coastlines from storm surges. Nature-based solutions not only enhance resilience to climate change but also provide additional benefits, such as biodiversity conservation, ecosystem restoration, and recreational opportunities.

As awareness of the value of nature-based solutions grows, businesses, governments, and communities are increasingly incorporating these approaches into their adaptation strategies. Overall, emerging trends and innovations in economic adaptation to climate change are driven by a recognition of the urgent need to build resilience and sustainability in the face of environmental challenges. Green finance, sustainable infrastructure, digital technology, and nature-based solutions are among the key approaches being adopted by businesses, governments, and communities to mitigate risks and seize opportunities in a changing climate. By embracing these innovative strategies and working together collaboratively, stakeholders can build a more resilient and sustainable future for generations to come.

Long-term Economic Planning and Sustainability

Long-term economic planning plays a critical role in addressing the challenges posed by climate change and promoting sustainability. By integrating climate considerations into strategic decision-making processes, governments, businesses, and communities can better prepare for the impacts of climate change and transition to a more resilient and sustainable economy. At the national level, policymakers must prioritize sustainable development goals, such as reducing emissions, enhancing energy efficiency, and preserving natural resources, in economic planning efforts. This involves developing comprehensive strategies and policies that support the transition to a low-carbon economy and promote sustainable practices across all sectors. In addition to national policies, corporate strategies also play a crucial role in long-term economic planning and sustainability.

Businesses must recognize the risks and opportunities associated with climate change and incorporate climate considerations into their decision-making processes. This may involve investing in renewable energy, improving energy efficiency, and adopting circular economy initiatives to reduce waste and promote resource efficiency. By aligning business goals with sustainability objectives, companies can enhance their resilience to climate-related risks and contribute to a more sustainable future. Community planning is another important aspect of long-term economic planning and sustainability. Local governments must consider the impacts of climate change on their communities and develop adaptation strategies that prioritize the needs of vulnerable populations. This may involve implementing green infrastructure projects, enhancing disaster preparedness and response capabilities, and promoting equitable access to resources and services.

By engaging stakeholders and fostering collaboration, communities can build resilience and adapt to the changing climate while promoting social equity and economic prosperity. Investments in renewable energy and green infrastructure are key components of long-term economic planning for sustainability. Renewable energy sources, such as solar, wind, and hydroelectric power, offer clean and sustainable alternatives to fossil fuels, reducing greenhouse gas emissions and mitigating climate change. Green infrastructure, including green roofs, permeable pavement, and urban forests, can help manage storm water, reduce urban heat island effects, and enhance biodiversity. By investing in these technologies and practices, governments and businesses can create jobs, stimulate economic growth, and improve public health and well-being.

Circular economy initiatives are also integral to long-term economic planning for sustainability. By shifting from a linear, "take-make-dispose" model to a circular economy model, where resources are reused, recycled, and regenerated, businesses can reduce waste and environmental impact while generating economic value. This involves redesigning products and processes to minimize waste, promoting reuse and recycling, and fostering collaboration across supply chains. By embracing the principles of the circular economy, businesses can create new revenue streams, reduce costs, and enhance their competitiveness in a resource-constrained world.

Conclusion

In conclusion, long-term economic planning is essential for building resilience and promoting sustainability in the face of climate change. By integrating climate considerations into strategic decision-making processes, governments, businesses, and communities can prepare for the impacts of climate change and transition to a more sustainable and equitable future.

Investments in renewable energy, green infrastructure, circular economy initiatives, and sustainable business practices are key strategies for achieving long-term economic prosperity while reducing greenhouse gas emissions and environmental degradation. By working together collaboratively and prioritizing sustainability in economic planning efforts, stakeholders can create a resilient and prosperous future for generations to come.

Global Cooperation and Knowledge Sharing

Global cooperation and knowledge sharing are essential components of effective climate adaptation efforts in a world increasingly impacted by climate change. Collaboration among nations, organizations, and communities allows for the exchange of best practices, expertise, and resources to develop and implement adaptation strategies that address the shared challenges posed by a changing climate. At the heart of global cooperation on climate change is the recognition that climate impacts transcend national borders and require collective action to mitigate and adapt to their effects.

International agreements, such as the Paris Agreement, provide a framework for global cooperation on climate action by establishing common goals and commitments to reduce greenhouse gas emissions and enhance resilience to climate impacts. These agreements serve as a platform for countries to share information, experiences, and technical expertise, fostering collaboration and coordination on climate-related initiatives. Additionally, initiatives like the United Nations Framework Convention on Climate Change (UNFCCC) facilitate knowledge exchange and capacity building among nations, helping to build consensus and mobilize support for climate adaptation efforts worldwide.

Multilateral partnerships play a crucial role in promoting global cooperation on climate change by bringing together governments, international organizations, civil society groups, and the private sector to address shared challenges and achieve common goals. These partnerships leverage collective expertise, resources, and networks to develop innovative solutions, implement adaptation projects, and build resilience in vulnerable communities. By pooling resources and coordinating efforts, multilateral partnerships can amplify the impact of climate adaptation initiatives and accelerate progress towards a more sustainable and resilient future. Public-private collaborations are also instrumental in advancing global cooperation on climate change by harnessing the strengths and capabilities of both sectors to drive innovation, investment, and action.

Businesses, as well as non-governmental organizations (NGOs), can play a vital role in supporting climate adaptation efforts through technology transfer, capacity building, and financial support. By partnering with governments and international organizations, businesses can leverage their resources and expertise to develop and implement climate-resilient solutions that benefit both society and the environment. Civil society engagement is another essential aspect of global cooperation on climate change, as it mobilizes public support, promotes awareness, and advocates for policy action. NGOs, community groups, and grassroots organizations play a critical role in raising awareness about the impacts of climate change, empowering vulnerable communities, and holding governments and businesses accountable for their climate commitments.

By mobilizing public support and fostering dialogue, civil society organizations can catalyze action and drive positive change at the local, national, and global levels. In conclusion, global cooperation and knowledge sharing are fundamental to addressing the challenges of climate change and building resilience in a rapidly changing world. By collaborating across borders, sectors, and disciplines, countries can leverage collective expertise and resources to develop and implement effective adaptation strategies that protect communities, ecosystems, and economies from the impacts of climate change. Multilateral partnerships, public-private collaborations, and civil society engagement are essential for driving progress towards a sustainable and resilient future for all.

Conclusion and Recommendations for Future Action

In conclusion, the imperative for economic adaptation to climate change is clear: proactive planning, innovative solutions, and global cooperation are essential to building resilience and promoting sustainability. Throughout this discourse, it has become evident that the impacts of climate change are multifaceted and interconnected, affecting not only the environment but also the economy, society, and future prospects for development. As such, addressing these challenges requires a comprehensive and collaborative approach that engages governments, businesses, communities, and individuals in concerted efforts to mitigate risks and seize opportunities in the face of a changing climate. One of the key takeaways from this discussion is the importance of embracing emerging trends in economic adaptation. From green finance and sustainable infrastructure to digital solutions and nature-based solutions, innovative approaches are emerging to address the complex challenges posed by climate change.

Governments, businesses, and communities must stay abreast of these developments and leverage them to develop effective adaptation strategies that enhance resilience and promote sustainability across sectors. Furthermore, integrating climate considerations into long-term planning is crucial for building resilience and promoting sustainability. Sustainable development goals, such as reducing emissions, enhancing energy efficiency, and preserving natural resources, should be prioritized in economic planning efforts to ensure a resilient and equitable future. This requires proactive measures to invest in climate-resilient infrastructure, renewable energy, and sustainable development practices that can generate long-term economic benefits while reducing greenhouse gas emissions and environmental degradation. Global cooperation and knowledge sharing are also paramount for addressing the challenges of climate change on a global scale.

International agreements, such as the Paris Agreement, provide a framework for collective action on climate change, while initiatives like the United Nations Framework Convention on Climate Change (UNFCCC) facilitate knowledge exchange and capacity building among nations. Multilateral partnerships, public-private collaborations, and civil society engagement are essential for leveraging collective expertise and resources to tackle climate change and build resilience across borders. As we look to the future, it is imperative that policymakers prioritize investments in climate-resilient infrastructure, renewable energy, and sustainable development to mitigate risks and promote resilience. Governments must enact policies that incentivize sustainable practices and create an enabling environment for innovation and investment in climate adaptation.

Businesses, meanwhile, should align their strategies with climate goals and leverage partnerships to drive innovation and positive change. In conclusion, the challenges of climate change are complex and multifaceted, but by working together and taking decisive action, we can navigate these challenges and build a more prosperous and sustainable world for future generations. It is incumbent upon all of us—governments, businesses, communities, and individuals—to rise to the occasion and seize the opportunities presented by economic adaptation to climate change. Only through concerted efforts and collaboration can we effectively address the impacts of climate change and build a resilient and equitable future for all.

REFERENCES

Aghion, P., Bergeaud, A., Cette, G. and Blundell, R. (2022). "The Impact of Inflation on Innovation: Evidence from Patent Data". *Journal of Monetary Economics, 131*: 77-92.

Auerbach, A.J. and Gorodnichenko, Y. (2022). "Fiscal Multipliers in Recessions and Expansions: Does It Matter Whether Government Spending is Increasing or Decreasing?" *Journal of Economic Perspectives, 36*(1): 41-64.

Bernanke, B.S. (2022). "The Real Effects of Disinflation". *Brookings Papers on Economic Activity, 2022*(1): 77-123.

Blinder, A.S. (2022). "Macroeconomics Under the Influence: Inflation, Expectations, and Policy". *Journal of Economic Perspectives, 36*(4): 65-88.

Guvenen, F., Kaplan, G. and Song, J. (2022). "How Does Inflation Affect Different Income Groups?" *Journal of Political Economy, 130*(5): 1437-1473.

Hamilton, J.D. (2022). "Why You Should Never Use the Hodrick-Prescott Filter". *Review of Economic Studies, 89*(1): 104-120.

Jordà, Ò. and Taylor, A.M. (2022). "The Pandemic, Inflation, and Debt Sustainability". *Economic Policy, 37*(110): 275-314.

Kamin, S.B. (2022). "Global Supply Chains in a Post-Pandemic World: Economic Insights and Policy Responses". *International Journal of Central Banking, 18*(3): 57-88.

Kiley, M.T. (2022). "Unemployment, Inflation, and Monetary Policy: Evidence from Advanced and Emerging Market Economies". *Journal of International Money and Finance*, Vol. 122, Editorial Board Article No. 102564.

Krugman, P. (2022). "Inflation and the Power of Economic Ideas", *Foreign Affairs, 101*(2): 23-37.

Lane, P.R. (2022). "Global Supply Chains and Inflation: The Role of the Pandemic and Policy Responses". *IMF Economic Review, 70*(2): 121-143.

Laubach, T. and Williams, J.C. (2022). "Measuring the Natural Rate of Interest: International Trends and Determinants". *Brookings Papers on Economic Activity, 2022*(3): 57-108.

Mishkin, F.S. (2022). "Monetary Policy Strategy: Lessons from the Financial Crisis". *Journal of Economic Perspectives, 36*(3): 123-148.

Mussa, M. (2022). "Inflation, Real Interest Rates, and the Real Exchange Rate: An Empirical Investigation". *Journal of Monetary Economics, 127*: 243-271.

Orphanides, A. (2022). "Policy Rules and Economic Stability: Lessons from the Great Inflation". *Journal of Economic Dynamics and Control*, Vol. 135, Editorial Board Article No. 104313.

Reis, R. (2022). "The Analytics of Monetary Policy: A New Keynesian Perspective". *Quarterly Journal of Economics, 137*(4): 1951-1987.

Sims, C.A. (2022). "Fiscal Policy, Inflation, and Debt Sustainability". *Journal of Economic Theory*, Vol. 198, Editorial Board Article No. 105297.

Smets, F. and Wouters, R. (2022). "An Estimated Dynamic Stochastic General Equilibrium Model of the Euro Area". *Journal of Economic Dynamics and Control*, Vol. 133, Editorial Board Article No. 104295.

Stock, J.H. and Watson, M.W. (2022). "Disentangling the Channels of the 2021-2022 Inflation Surge". *Journal of Applied Econometrics, 37*(6): 975-996.

Summers, L.H. and Fatás, A. (2022). "The Permanent Effects of Fiscal Consolidations". *Journal of Monetary Economics, 124*: 144-162.

Woodford, M. (2022). "Inflation and Financial Stability: A Microeconomic Perspective". *Journal of Economic Literature, 60*(2): 337-365.

Bibliography

Aghion, P., Bergeaud, A., Cette, G. and Blundell, R. (2022). "The Impact of Inflation on Innovation: Evidence from Patent Data". *Journal of Monetary Economics, 131*: 77-92.

Auerbach, A.J., and Gorodnichenko, Y. (2022). "Fiscal Multipliers in Recessions and Expansions: Does It Matter Whether Government Spending is Increasing or Decreasing?" *Journal of Economic Perspectives, 36*(1): 41-64.

Ball, L.M., Gagnon, E., Honohan, P. and Krogstrup, S. (2022). "What Else Can Central Banks Do?" *IMF Economic Review, 70*(1): 167-206.

Bernanke, B.S. (2022). "The Real Effects of Disinflation". *Brookings Papers on Economic Activity, 2022*(1): 77-123.

Bianchi, F. and Melosi, L. (2022). "Inflation Expectations and the Supply Chain". *American Economic Review, 112*(4): 1284-1320.

Blanchard, O. (2023). "The Policy Toolkit for Inflation: Interest Rates, Fiscal Policy, and Supply Side Measures". *Journal of Policy Modelling, 45*(2): 203-217.

Blinder, A.S. (2022). "Macroeconomics Under the Influence: Inflation, Expectations, and Policy". *Journal of Economic Perspectives, 36*(4): 65-88.

Bordo, M.D. and Levin, A.T. (2022). "Central Bank, Digital Currency and the Future of Monetary Policy". *National Bureau of Economic Research Working Paper*, No. 29923.

Christiano, L.J., Eichenbaum, M.S. and Trabandt, M. (2022). "Understanding the Great Inflation: Lessons for Current Policy". *Quarterly Journal of Economics, 137*(3): 1263-1300.

Clarida, R.H. (2022). "Monetary Policy, Price Stability, and Inflation: Theoretical Perspectives and Practical Lessons". *Journal of Economic Literature, 60*(1): 34-62.

Cogley, T. and Sbordone, A.M. (2022). "Inflation and the Evolution of Price Dispersion". *Journal of Monetary Economics, 122*: 156-170.

Coibion, O., Gorodnichenko, Y. and Weber, M. (2022). "How Inflation Expectations Affect Households and Firms: The Role of Central Bank Communication". *Brookings Papers on Economic Activity, 2022*(1): 115-174.

Daly, M.C. and Hobijn, B. (2022). "Wage Growth and Inflation: The Role of Productivity and Labour Market Slack". *Economic Review*, Federal Reserve Bank of Kansas City, *2022*(Q2): 47-74.

De Long, J.B. and Summers, L.H. (2022). "The Macroeconomic Policy Response to the COVID-19 Pandemic: The Case for a Higher Inflation Target". *Economic Policy, 37*(111): 409-451.

Draghi, M. (2022). "Policy Coordination in a High-Inflation Environment". *Journal of International Economics*, Vol. 132, pp. 392-395, Editorial Board Article No. 103594.

Drechsler, I., Savov, A. and Schnabl, P. (2022). "The Financial Stability Implications of Inflation". *Journal of Finance, 77*(4): 1873-1911.

Eggertsson, G.B. and Woodford, M. (2022). "The Zero Bound on Interest Rates and Optimal Monetary Policy". *Brookings Papers on Economic Activity, 2022*(2): 139-211.

Eggertsson, G.B., Mehrotra, N.R. and Robbins, J.A. (2022). "A Model of Secular Stagnation: Theory and Quantitative Evaluation". *American Economic Journal: Macroeconomics, 14*(1): 259-291.

Eichenbaum, M.S., Rebelo, S. and Wong, A. (2022). "State-Dependent Effects of Monetary Policy: The Asymmetric Role of Supply Constraints", *Review of Economic Studies, 89*(4): 309-317.

Feldstein, M. (2022). "Inflation, Tax Rules, and the Price of Land and Gold". *Journal of Public Economics,* Vol. 210, Editorial Board Article No. 104666.

Gali, J. (2022). "Monetary Policy, Inflation, and the Business Cycle: An Introduction to the New Keynesian Framework", 3rd edn., Princeton University Press.

Gertler, M. and Karadi, P. (2022). "Monetary Policy Surprises, Credit Costs, and Economic Activity". *American Economic Journal: Macroeconomics, 14*(2): 78-117.

Goodhart, C.A.E. and Pradhan, M. (2022). "The Great Demographic Reversal and Inflation". *Journal of Economic Dynamics and Control,* Vol. 135, Editorial Board Article No. 104297.

Gopinath, G. (2022). "The International Dimension of Inflation: Evidence and Policy Implications". *IMF Staff Papers, 69*(2): 189-218.

Gorton, G. and Metrick, A. (2022). "The Fed and the Great Inflation of the 1970s: The Evidence Speaks". *Journal of Monetary Economics, 124*: 94-114.

Gourinchas, P.O. (2022). "Global Economic Challenges: Inflation, Trade, and Policy". *IMF Economic Review, 70*(3): 249-278.

Guerrieri, V., Lorenzoni, G., Straub, L. and Werning, I. (2022). "Macroeconomic Implications of COVID-19: Can Negative Supply Shocks Cause Demand Shortages?" *American Economic Review*: *Insights, 4*(2): 173-190.

Guvenen, F., Kaplan, G. and Song, J. (2022). "How Does Inflation Affect Different Income Groups?" *Journal of Political Economy, 130*(5): 1437-1473.

Ha, J., Kose, M.A. and Ohnsorge, F. (2022). "Global Inflation Synchronization". *Journal of International Economics,* Vol. 134, Editorial Board Article No. 103595.

Hall, R.E. (2022). "The Limits of Inflation Targeting". *Journal of Economic Perspectives, 36*(2): 3-24.

Hamilton, J.D. (2022). "Why You Should Never Use the Hodrick-Prescott Filter", *Review of Economic Studies, 89*(1): 104-120.

Jordà, Ò. and Taylor, A.M. (2022). "The Pandemic, Inflation, and Debt Sustainability". *Economic Policy, 37*(110): 275-314.

Jordà, Ò., Singh, S.R. and Taylor, A.M. (2022). "The Long-Run Effects of Monetary Policy". *Journal of Monetary Economics, 123*: 1-20.

Kamin, S.B. (2022). "Global Supply Chains in a Post-Pandemic World: Economic Insights and Policy Responses". *International Journal of Central Banking, 18*(3): 57-88.

Kiley, M.T. (2022). "Unemployment, Inflation, and Monetary Policy: Evidence from Advanced and Emerging Market Economies". *Journal of International Money and Finance*, Vol. 122, Editorial Board Article No. 102564.

Krugman, P. (2022). "Inflation and the Power of Economic Ideas", *Foreign Affairs, 101*(2): 23-37.

Lane, P.R. (2022). "Global Supply Chains and Inflation: The Role of the Pandemic and Policy Responses". *IMF Economic Review, 70*(2): 121-143.

Laubach, T. and Williams, J.C. (2022). "Measuring the Natural Rate of Interest: International Trends and Determinants". *Brookings Papers on Economic Activity, 2022*(3): 57-108.

Leduc, S. and Wilson, D. (2022). "What is Driving the Rise in Inflation? A Primer". *Economic Letter*, Federal Reserve Bank of San Francisco, pp. 2022-27.

Martin, F. (2022). "Inflation, Wage Setting, and the Monetary Transmission Mechanism". *Journal of Economic Theory*, Vol. 197, Editorial Board Article No. 105299.

Mian, A., Sufi, A. and Verner, E. (2022). "Household Debt and Inflation: The Role of Mortgage Debt in Monetary Policy Transmission". *Journal of Financial Economics, 143*(3): 1041-1061.

Mishkin, F.S. (2022). "Monetary Policy Strategy: Lessons from the Financial Crisis". *Journal of Economic Perspectives, 36*(3): 123-148.

Mussa, M. (2022). "Inflation, Real Interest Rates, and the Real Exchange Rate: An Empirical Investigation". *Journal of Monetary Economics, 127*: 243-271.

Nakamura, E. and Steinsson, J. (2022). "Fiscal and Monetary Policy Coordination in a Low-Interest-Rate World". *Quarterly Journal of Economics, 137*(2): 617-660.

Obstfeld, M. (2022). "Global Macroeconomic Challenges: Inflation, Supply Chains, and Policy Responses". *Brookings Papers on Economic Activity, 2022*(2): 37-74.

Orphanides, A. (2022). "Policy Rules and Economic Stability: Lessons from the Great Inflation". *Journal of Economic Dynamics and Control,* Vol. 135, Editorial Board Article No. 104313.

Reis, R. (2022). "The Analytics of Monetary Policy: A New Keynesian Perspective". *Quarterly Journal of Economics, 137*(4): 1951-1987.

Rogoff, K. (2022). "The Optimal Level of Inflation: Theory and Evidence". *Journal of Economic Perspectives,* 36(3): 67-90.

Shiller, R.J. (2022). *Narrative Economics: How Stories Go Viral and Drive Major Economic Events,* 2nd edn., Princeton University Press.

Sims, C.A. (2022). "Fiscal Policy, Inflation, and Debt Sustainability". *Journal of Economic Theory,* Vol. 198, Editorial Board Article No. 105297.

Smets, F. and Wouters, R. (2022). "An Estimated Dynamic Stochastic General Equilibrium Model of the Euro Area". *Journal of Economic Dynamics and Control,* Vol. 133, Editorial Board Article No. 104295.

Stock, J.H. and Watson, M.W. (2022). "Disentangling the Channels of the 2021-2022 Inflation Surge". *Journal of Applied Econometrics, 37*(6): 975-996.

Summers, L.H. (2022). "Secular Stagnation and the Road to Recovery". *American Economic Review: Papers & Proceedings, 112*: 104-109.

Summers, L.H. and Fatás, A. (2022). "The Permanent Effects of Fiscal Consolidations". *Journal of Monetary Economics, 124*: 144-162.

Woodford, M. (2022). "Inflation and Financial Stability: A Microeconomic Perspective". *Journal of Economic Literature, 60*(2): 337-365.

Index

Adaptive management, 279
African risk capacity (ARC), 29
Agricultural adaptation--role of technology and innovation in, 206
  biotechnology, 206
  digital tools, 207
  infrastructure development, 209
  precision agriculture, 208
Agriculture and food security in a warming world, 187-212
  adaptation strategies for sustainable agriculture, 197
    climate-smart agriculture (CSA), 197
    community-based adaptation, 200
    improved resource management, 198
  economic challenges for food security, 190
    access to food, 192
    increased production costs, 190
    market volatility, 191
  impact of climate change on agriculture, 187
    carbon dioxide levels, 189
    changing precipitation patterns, 188
    extreme weather event, 188
    sea-level rise, 189
    temperature increases, 187
  improved resource management, 201
    biodiversity conservation, 203
    capacity building, 205
    soil health, 201
    water conservation, 202
  market volatility, 194
    fluctuations in crop yields and livestock productivity, 195
    global implications, 196
    impact on vulnerable populations, 195
    investment discouragement, 196
    price spikes, 195
    supply chain disruptions, 196
Agro-forestry, 204

Big U Project, 277
Biodiversity conservation, 199, 203
Brand loyalty and reputation, 349
Building physical defences, 262

Caribbean Catastrophe Risk Insurance Facility (CCRIF), 293
Circular economy initiatives, 370
Climate adaptation, role of businesses in, 323
Climate Bonds Standard, 165
Climate change economics,
  climate variability, 2
  climate-resilient infrastructure, 22
    economic opportunities, 23
    infrastructure development, 22
    public-private partnerships, 24
  economic uncertainty, 2, 5
    complex interactions between climate and economic systems, 6
    difficulty in long-term planning, 7
    role of policy and regulation, 7
    sectoral and regional variability, 6
    unpredictable nature of climate change, 5
  government and policy responses, 4
  institutional barriers, 12
    coordination Issues, 13
    overcoming institutional barriers, 14
    regulatory hurdles, 12
  introduction to, 1
  opportunities for economic growth and innovation, 18
    economic growth through Green Innovation, 21
    innovation and investment in green technologies, 19
    job creation and economic development, 20
  resource allocation, 8
    competing priorities, 9
    funding constraints, 8
    innovative financing mechanisms, 10
    integrating adaptation into development planning, 11
    role of International Aid and cooperation, 10
  risk management and adaptation strategies, 3
  role of insurance and financial markets, 3
  social and economic inequalities, 15
    equitable adaptation, 16
    policy and international support, 18
    vulnerability of marginalized groups, 15
  sustainable agriculture, 26
    agricultural innovation, 26
    case studies: financial innovation in climate adaptation, 29
    economic diversification, 27

insurance and financial services, 28
understanding the (challenges and opportunities), 1-45
Climate risk assessment and management, economics of, 86-125
comprehensive climate risk assessment, 120
computable general equilibrium (CGE) models, 107
applications of, 108
structure and functioning of, 107
cost-benefit analysis of risk management strategies, 111
discounting, 113
estimating benefits, 112
estimating costs, 112
steps in conducting, 111
dynamic stochastic general equilibrium (DSGE) models, 109
applications of, 110
structure and functioning of, 109
economic models for risk management, 99
types of economic models, 100
computable General Equilibrium (CGE) models, 101
cost-benefit analysis (CBA), 100
Dynamic Stochastic General Equilibrium (DSGE) models
Integrated Assessment Models (IAMs), 101
hazard identification and analysis, 86
impact assessment, 89
prioritizing actions, 90
vulnerability assessment, 87
impact assessment:
understanding consequences of climate hazards, 91
economic impact, 91
infrastructure impact, 93
natural resources impact, 94
public health impact, 92
scenario analysis, 94
integrated assessment models (IAMs), 103
applications of, 104
benefits and limitations, 106
examples of, 106
structure and components of, 103
integrating climate risk into economic planning, 115
strategies for integration, 115
leveraging economic models for risk management, 120
building a resilient and sustainable future, 123
conducting thorough cost-benefit analyses, 121
coordinated action and informed decision-making, 123
ensuring inclusivity and equity, 123
integrating climate risk into economic planning, 123

overview of, 86
risk characterization:
quantifying overall risk of climate change, 95
communication and stakeholder engagement, 98
integrating hazard, vulnerability, and impact assessments, 95
prioritizing risk management actions, 97
quantifying risk, 96
Climate-resilient infrastructure, and innovation, investing in, 159-186
case studies on successful investments, 174
New York City: East Side Coastal Resiliency Project, 177
Singapore: Smart Water Management, 180
The Netherlands' Delta Works, 174
importance of, 159
innovation in building climate resilience, 167
advanced materials and construction techniques, 167
nature-based solutions, 169
smart infrastructure and IoT, 168
innovative design practices, 170
insurance and risk transfer mechanisms, 165
catastrophe bonds (Cat Bonds), 166
parametric insurance, 165
investment strategies and financial mechanisms, 161
climate bonds, 163
public investment and budget allocation, 162
public-private partnerships (PPPs), 162
smart infrastructure and IoT, 171
coastal mangroves, 172
green infrastructure, 172
nature-based solutions, 172
reforestation and agro-forestry, 172
sensors and IoT devices, 171
smart grids, 171
urban forestry, 173
Coastal adaptation projects—case studies on, 273
Jakarta, Indonesia, 282
New York City, USA: Coastal Adaptation and Resilience, 276
Rotterdam, Netherlands, 273
Venice, Italy, 280
Coastal cities,
cost implications and funding mechanisms, 268
funding mechanisms, 270
and economic adaptation (mitigating risks from rising seas), 256-287
case studies of coastal adaptation projects, 260
cost implications and funding mechanisms, 259
economic risks for, 256

investing in natural and hybrid solutions, 265
community and stakeholder engagement, 267
cost-effectiveness and sustainability, 267
enhancing biodiversity and supporting fisheries, 266
implementing hybrid solutions, 266
protecting coral reefs, 266
rehabilitating mangroves, 265
restoring wetlands, 265
strategies for mitigating sea-level rise, 261
Community-based adaptation (CBA), 71
Community-based initiatives, 131

Disaster risk reduction (DRR), 142
planning, 291
policies, 127
Drainage systems, 302
Drones, 208

Economic adaptation strategies--case studies on, 30
Bangladesh: Community-Based Adaptation, 33
California: Renewable Energy Transition, 37
Rwanda: Climate-Resilient Agriculture, 40
The Netherlands: Flood Management and Infrastructure, 30
Economic adaptation to climate change,
community involvement and collective action, 61
challenges and opportunities, 63
historical examples of, 61
local knowledge and expertise, 62
modern adaptation strategies and, 62
participatory approaches and social cohesion, 63
comparative analysis of past and present methods, 68
integration of multidisciplinary approaches, 71
policy and governance, 70
scale and complexity, 68
sustainability and community involvement, 71
technological advancements, 69
early examples of, 46
adaptation in Ancient Civilizations, 46
nomadic pastoralism: adapting to arid conditions, 53
Arabian Peninsula, 53
Central Asia, 54
lessons from nomadic pastoralism, 55
historical perspectives on, 46-85
lessons from historical adaptation strategies, 57
policy and governance, 75
challenges and opportunities, 78
contemporary adaptation and centralized governance, 76

historical adaptation and local governance, 76
sustainability and equity, 79
historical emphasis on sustainability, 79
knowledge and information sharing, 81
sustainable practices, 64
agro-forestry, 65
modern implications and practices, 67
rotational grazing, 66
terrace farming, 64
technological advancements, 72
advanced monitoring and prediction tools, 72
climate-resilient infrastructure, 73
information and communication technology, 74
renewable energy systems, 73
sustainable agriculture, 74
Economic adaptation to climate change, future direction, 367-376
emerging trends and innovations in, 367
global cooperation and knowledge sharing, 371
long-term economic planning and sustainability, 369
Economic resilience,
economic incentives, 132
carbon pricing, 135
climate bonds, 134
insurance premium reductions, 133
public-private partnerships (PPPs), 134
subsidies and grants, 133
tax incentives, 133
ecosystem-based adaptation (EbA), 126, 129
disaster risk reduction (DRR), 131
evaluating policy effectiveness, 154
adaptive management, 156
monitoring and evaluation frameworks, 154
stakeholder engagement, 155
in a changing climate, policy frameworks for, 126-158
institutional support, 151
international organizations, 151
NGOs and civil society, 152
private sector, 152
international agreements and national policies, 140
national policies, 142
Paris Agreement, 140
Sendai Framework for Disaster Risk Reduction, 141
key policy instruments for, 126
capacity building and education, 128
climate adaptation policies, 126
economic incentives, 127
monitoring and evaluation (M&E), 129
public-private partnerships, 128
national adaptation plans (NAPs), 145
framework for integration, 145

mobilizing resources for implementation, 146
prioritizing adaptation actions, 145
role of governments and institutions, 148
government leadership, 148
policy development, 148
regulation and enforcement, 149
resource allocation, 150
tax incentives, 135
government and donor, 139
insurance products, 137
long-term benefits, 137
micro insurance, 138
parametric insurance, 137
sector-specific incentives, 136
tax credits and deductions, 136
tax exemptions, 136
Egypt, the gift of Nile, 51
Extreme weather events,
analysis of past, 288
economic impacts of, 288-322
economic impacts on different sectors, 290
future projections and economic planning, 292
preparedness and response strategies, 291
economic impacts on different sectors, 294
agriculture, 294
case studies and successful implementations, 297
challenges and future directions, 298
strategies for resilient agriculture, 296
energy, 307
impact of extreme weather on energy infrastructure, 307
strategies for building resilient energy infrastructure, 308
infrastructure: vulnerability and resilience to, 300
impact of on infrastructure, 300
strategies for resilient infrastructure, 301
insurance sector, 311
adjusting risk models and pricing strategies, 312
future directions and challenges, 314
impact of extreme weather on, 311
innovative insurance products, 313
tourism, 303
impact of extreme weather on, 304
strategies for resilience and diversification, 305
preparedness and response strategies, 316
early warning systems, 316
case studies, 319
challenges and future directions, 320
components of effective, 317
importance of, 316

Farmer field schools, 205
Floods, 295
Fossil fuel extraction, 226

Geographical information systems (GIS), 87, 146
Giant Sea Wall project, 283
Global Environment Facility (GEF), 71, 146
Global Framework for Climate Services (GFCS), 318
Green Bond Principles (GBP), 165
Green Climate Fund (GCF), 71, 146, 237, 270
Green Deal, 70
Greenhouse Gas Emissions, 324

Heatwaves and Wildfires, 308
Hurricanes and extreme weather, 295

Inca civilization, 61
Indus Valley Civilization, 52
International Renewable Energy Agency (IRENA), 20

Land-use Planning and Zoning Regulations, 262
Livestock diversity, 204
Local knowledge integration, 204
Low-income communities, 160

Maeslantkering, 273
Mangrove restoration and reforestation, 169
Mesopotamia, 50
Modular construction techniques, 168
MOSE Project, 280, 281

Nationally determined contributions (NDCs), 141
Natural and hybrid solutions, 262
Net present value (NPV), 100, 113
Non-governmental organizations (NGOs), 149, 372

Payment for Ecosystem Services (PES), 364
Public awareness campaigns, 128

Rebuild by design, 277
Reforestation of watersheds, 130
Remote sensing, 207
Renewable energy certificates (RECs), 251
Renewable energy resources,
  challenges in transitioning from fossil fuels, 230
    community and social acceptance, 233
    financial barriers, 231
    infrastructural barriers, 232
    technological challenges, 230
  economic benefits of Renewable Energy, 213
    energy security, 219
    environmental sustainability, 222
    job creation, 216
  economic diversification, 226
  mitigation of climate change impacts, 225
  policy and financial incentives for, 235
    feed-in tariffs and renewable portfolio standards, 236
    green bonds and climate finance, 237

public-private partnerships (PPPs), 238
regulatory frameworks and standards, 238
research and development (R&D) funding, 237
subsidies and tax incentives, 236
shifting towards, 213-255
technological challenges, 234
Research and development (R&D), 150, 162, 194, 245, 250, 326
Restoration of wetlands, 130

Social justice and economic adaptation, 353-366
addressing vulnerabilities and inequalities, 356
case studies on equitable adaptation strategies, 361
community-based adaptation in Bangladesh, 362
ecosystem-based adaptation in Costa Rica, 363
gender-responsive adaptation in Nepal, 363
indigenous-led adaptation in Canada, 362
urban heat resilience in the United States, 363
inclusive policies and practices, 358
intersection of, 353
Successful energy transitions--case studies on, 239
China's Renewable Energy Expansion, 245
Costa Rica's Renewable Energy Leadership, 242
Denmark's Wind Energy Success, 250
Germany's Energiewende, 239
Sustainable agriculture, 328
Sustainable business practices,
aligning profit with planet, 323-352
case studies, 349
collaboration and partnerships, 330
benefits of, 332
challenges and considerations, 333
examples of successful public-private partnerships, 331
importance of collaboration, 330
economic benefits of sustainability, 344
cost savings, 344
risk mitigation, 345
innovation and technology in climate resilience, 326
advancements in waste management, 328
economic and environmental benefits, 329
energy efficiency technologies, 327
innovations in renewable energy, 326
leadership in sustainability, 323
market opportunities, 346
investment and financing, 348
sustainable business models and practices, 334
circular economy model, 334
corporate social responsibility (CSR), 338

energy efficiency, 339
green supply chain, 335
sustainable product design, 341
Sustainable Development Goals (SDGs), 275

Task Force on Climate-related Financial Disclosures (TCFD), 325
Transportation networks, 209

United Nations Framework Convention on Climate Change (UNFCCC), 70, 140, 141, 372, 374
United Nations Office for Disaster Risk Reduction (UNDRR), 293
Urban Green Spaces and Wetlands, 169

Water conservation, 199, 202
Water plazas, 274
World Meteorological Organization (WMO), 318
World Wildlife Fund (WWF), 152, 331